JMP® for Mixed Models

Ruth M. Hummel
Elizabeth A. Claassen
Russell D. Wolfinger

sas.com/books

The correct bibliographic citation for this manual is as follows: Hummel, Ruth M., Elizabeth A. Claassen, and Russell D. Wolfinger. 2021. *JMP® for Mixed Models*. Cary, NC: SAS Institute Inc.

JMP® for Mixed Models

ISBN 978-1-952365-21-8 (Hardcover)
ISBN 978-1-951684-02-0 (Paperback)
ISBN 978-1-951684-03-7 (Web PDF)
ISBN 978-1-952363-85-6 (EPUB)
ISBN 978-1-952363-86-3 (Kindle)

SAS Institute Inc., SAS Campus Drive, Cary, NC 27513-2414

June 2021

Contents

About This Book

What Does This Book Cover?

Mixed models, which are an extension of classic statistical linear models (including analysis-of-variance and regression), are one of the most powerful and useful collection of methods for analyzing data from designed experiments. Variations of mixed models have been one of the strongest capabilities of SAS software since its beginnings in the mid 1970s. In parallel, JMP (a SAS product launched in 1989) has evolved into an incredibly powerful and popular tool for scientists and engineers. This book brings together these two legacies, and in example-driven fashion, walks through the core concepts of mixed models and how to best apply them in practice.

Mixed models are largely about how to handle experimental observations that are correlated. After introducing foundational concepts and terminology of mixed models with examples, the book covers increasing levels of complexity, revealing the richness and wide applicability of mixed models in most any discipline that collects data in well-formed experiments.

The first four chapters focus on mixed models in the context of analysis-of-variance (ANOVA). We find that ANOVA is a good place to start as it helps organize thinking around factors with a discrete number of levels, as are nearly always found in designed experiments. We proceed further and utilize a helpful construct known as a Skeleton ANOVA to help clearly break down and understand degrees of freedom as well as how information from experimental units in the design is being allocated to effects in the model.

Chapters 5 and 6 shift focus to continuous effects as commonly found in linear regression and repeated measures contexts. They fit quite naturally into the mixed model framework and can be effectively combined with ANOVA-style effects to handle a wide variety of common experimental setups. Chapter 7 covers spatial models, which extend mixed models further to handle covariance over two or more dimensions.

Chapter 8 shows how you can use simulation to rigorously explore deeper statistical properties of mixed models such as power and sampling distributions of outputs. Chapter 9 provides an introduction to generalized linear mixed models, which are used when the response is no longer normally distributed, such as when it is a discrete number of successes or a count. Chapter 10 concludes the book with discussions on how mixed models relate to current controversies in the statistical and broader scientific and engineering communities.

Is This Book For You?

JMP for Mixed Models builds on the success of the *SAS for Mixed Models* book series as well as several other related books and articles. In contrast to the SAS procedures and code forming the basis of previous books, JMP and JMP Pro offer the ability to fit mixed models from a dynamically interactive, mouse-driven interface. This enables you to use mixed models without having to write code and get to important results faster and with less effort. This book is designed as an instructional guide along these lines and is the very first of its kind in this regard.

If you fit one of the following two characterizations, this book is likely for you:

- You are a scientist or engineer running experiments in which subsets of the observations are correlated due to the design or the nature of the experimental units themselves. This includes designs such as a randomized block or split-plot as generated by JMP's rich design-of-experiments (DOE) routines.
- You are familiar with running mixed models, hierarchical linear models, or multilevel models in SAS, R, or other languages and want to learn an easy, point-and-click interface to fit them and obtain dynamically integrated statistics and graphics to aid in their interpretation and presentation.

If you take the time to learn mixed models in JMP, they will likely become one of the most useful tools that you have for analyzing designed experiments.

What Are the Prerequisites for This Book?

We assume you have knowledge of introductory statistical concepts such as those taught in an advanced high school or first-year college curriculum. This includes topics such as the following:

- Statistical Testing (of the mean, of the difference between two means), standard errors (of the mean, difference between two means), and t tests
- Distributions (normal, binomial, uniform, t, chi-square, F)
- One-Way ANOVA
- Factorial ANOVA
- Regression
- ANCOVA (regression with groups)

In JMP or JMP Pro, the Fit Model platform is the central one we will use, and some basic familiarity with it will be very helpful.

What Should You Know about the Examples?

Each topical chapter in this book begins with a description of several motivating examples that utilize the topic, and then we present the necessary conceptual background. With the background in place, we analyze the examples using both JMP and JMP Pro, including full interpretations of the output.

If you already have a decent understanding of mixed models and/or JMP, you may want to skip straight to examples that best match the problem that you want to analyze. Although the book roughly proceeds from simpler to more complex topics in a somewhat logical fashion, it is also designed to be a reference book in which you can find an example that most closely matches your current problem and skip directly to it.

Software Used to Develop the Book's Content

We use both JMP and JMP Pro throughout, highlighting key differences as they arise.

Example Code and Data

JMP tables for all of the examples are available in the books supplemental information web page. JMP Scripting Language (JSL) programs are either included with the tables themselves or provided as stand-alone programs that you can open in JMP and run.

Output and Graphics

All of the books output and graphics are generated on an Apple MacIntosh. If you are running on Microsoft Windows, the aesthetics of the output will be somewhat different but content should be the same.

We use several typeface conventions throughout the book to help demarcate between data set names, variable names, commands, etc. `Data sets`, `variables`, and `functions` are monospace. *Platforms, menus, options, variable roles, buttons,* and *function groups,* basically anything you click on, are italicized. Bold font is mostly used only for section headings, but it is also used to call out **table names** in JMP reports.

We Want to Hear from You

SAS Press books are written by SAS Users for SAS Users. We welcome your participation in their development and your feedback on SAS Press books that you are using. Please visit `sas.com/books` to do the following:

Sign up to review a book
Recommend a topic
Request information on how to become a SAS Press author
Provide feedback on a book

Do you have questions about a SAS Press book that you are reading? Contact the author through saspress@sas.com or `support.sas.com/author_feedback`. SAS has many resources to help you find answers and expand your knowledge. If you need additional help, see our list of resources: `sas.com/books`.

About the Authors

Ruth M. Hummel, PhD, is a Senior Manager of Analytical Education at SAS. Dr. Hummel develops curricula, teaches, and consults to help researchers and practitioners apply statistical methods and analytics to solving problems, predominantly in the health and life sciences. Prior to joining SAS in 2016, she worked at the Environmental Protection Agency as the statistical expert for the Risk Assessment Division of the Office of Pollution Prevention and Toxics, and she taught and consulted at the Pennsylvania State University and at the University of Florida. Dr. Hummel is a co-author of Business Statistics and Analytics in Practice, 9th edition, a business statistics textbook emphasizing simple data mining techniques earlier in the standard curriculum. She has a PhD in statistics from the Pennsylvania State University.

Elizabeth A. Claassen, PhD, is Research Statistician Developer in the JMP division of SAS. Dr. Claassen has over ten years of experience with SAS software and over eight years of experience with JMP. Her chief interest is generalized linear mixed models, and she brings to this work her expertise from testing and developing the mixed modeling capabilities in JMP as well as experience with SAS procedures for linear models. Dr. Claassen earned an MS and PhD in statistics from the University of Nebraska–Lincoln, where she received the Holling Family Award for Teaching Excellence from the College of Agricultural Sciences and Natural Resources.

Russell D. Wolfinger, PhD, is Director of Scientific Discovery and Genomics at SAS, a department that he started, where he leads research and development of JMP and SAS software solutions in the domains of genomics and clinical trials. Dr. Wolfinger devoted 10 years to developing and promoting SAS statistical procedures for mixed models and multiple testing. He has co-authored more than 100 publications, including three books. He is a Kaggle Grandmaster, as well as a Fellow of both the American Association for the Advancement of Science and the American Statistical Association.

Learn more about these authors by visiting their author pages, where you can download free book excerpts, access example code and data, read the latest reviews, get updates, and more:

http://support.sas.com/hummel
http://support.sas.com/claassen
http://support.sas.com/wolfinger

Acknowledgements

We first wish to thank John Sall, co-founder of SAS and creator of JMP, for his vision for the amazing product that is JMP software. We stand on the shoulders of giants. John is one we are proud to work with every day.

We also acknowledge the incredible contributions of Chris Gotwalt in the implementation of most of the Mixed Model routines used in this book.

We are deeply indebted to this book's 'forefathers', the *SAS for Mixed Models* series. Ramon Littell, George Milliken, Walter Stroup, and Oliver Schabenberger all left their imprint upon the mixed model community through their authorship, educational outreach, and support of modelers around the world.

We extend special thanks to the entire editorial and production teams at SAS Press. Our editor, Catherine Connolly, worked diligently to keep us on track even as the additional stresses of COVID-19 caused small delays. She also stepped in during a time of internal restructuring and was our copyeditor, improving the quality of this book in more ways.

Thank you to the reviewers of this manuscript. Your questions helped us clarify. Your expertise made us better. Your willingness to assist in the creation of this book puts us in your debt.

Special thanks to Ian Kleinfeld who graciously turned our vague vision for the cover art for this book into a reality on short, late notice. You practically read our minds.

Finally, we extend a most special appreciation to our families for their love and support while working on this book. To our extended JMP and SAS family, we offer sincere and deep gratitude. We could not ask for a better, more intelligent and fun group of colleagues, who have contributed in countless ways to the great software on which this book is based.

Chapter 1

Introduction

1.1 What is a Mixed Model?

Imagine you are lab scientist studying the effect of two chemicals, A and B, on cell viability. You prepare nine plates of media with healthy cells growing on each, and then apply A and B to randomly assigned halves of each plate. After a suitable incubation period, you collect treated cells from the halves of each plate and perform an assay on each sample to compute a measurement Y of interest. Four of the samples are accidentally contaminated during processing and produce no assay results. Your data table in JMP looks like Figure 1.1.

How should you analyze these data? A primary goal is to estimate the causal effect of Chemical on Y, while taking appropriate account of the experiment design based on Plate. A standard way to begin is to formulate a statistical model of Y as a function of Chemical and Plate. A *statistical model* is a mathematical equation formed using parameters and probability distributions to approximate a data-generating process. We refer to Y as the *response* in the model, or alternatively as the *dependent variable* or *target*. We refer to Chemical and Plate as *factors* or *independent variables*.

Note the different natures of Chemical and Plate. Chemical has two specifically chosen levels, A and B, whereas the levels of Plate are effectively a random set of such plates you routinely make in your lab. This is a most basic example of a case in which you would want to use a *mixed model*, which is a statistical model that includes both *fixed effects* and *random effects*. Here Chemical would be considered a fixed effect and Plate a random effect.

Figure 1.1: Cell Viability Data

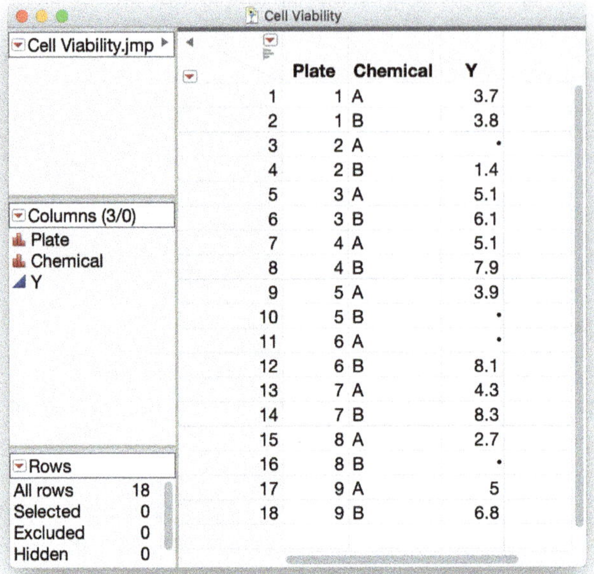

Cell Viability

Cell Viability.jmp

Columns (3/0)
Plate
Chemical
Y

Rows
All rows 18
Selected 0
Excluded 0
Hidden 0

	Plate	Chemical	Y
1	1	A	3.7
2	1	B	3.8
3	2	A	•
4	2	B	1.4
5	3	A	5.1
6	3	B	6.1
7	4	A	5.1
8	4	B	7.9
9	5	A	3.9
10	5	B	•
11	6	A	•
12	6	B	8.1
13	7	A	4.3
14	7	B	8.3
15	8	A	2.7
16	8	B	•
17	9	A	5
18	9	B	6.8

Key Terminology

Fixed Effect A statistical modeling factor whose specific levels and associated parameters are assumed to be constant in the experiment and across a population of interest. Scientific interest focuses on these specific levels. For example, when modeling results from three possible treatments, your focus is on which of the three is best and how they differ from each other.

Random Effect A statistical modeling factor whose observed values are assumed to arise from a probability distribution, typically assumed to be normal (Gaussian). Random effects can be viewed as a random sample from a population that forms part of the process that generates the data you observe. You want to learn about characteristics of the population and how it drives variability and correlations in your data. You want inferences about fixed effects in the same model to apply to the population corresponding to this random effect. You may also want to estimate or predict the realized values of the random effects.

Mixed Model A statistical model that includes both fixed effects and random effects.

Why is the distinction between fixed and random effects important? Many, if not most, real-life data sets do not satisfy the standard statistical assumption of independent observations. In the example above, we naturally expect observations from the same plate to be correlated as opposed to those from different plates. Random effects provide an easy and effective way to directly model this correlation and thereby enable more accurate inferences about other effects in the model. In the example, specifying Plate as a random effect enables us to draw better inferences about Chemical. Failure to appropriately model design structure such as this can easily result in biased inferences. With an appropriate mixed model, we can estimate primary effects of interest as well as compare sources of variability using common forms of dependence among sets of observations.

The use of fixed and random effects have a rich history, with countless successful applications in most every major scientific discipline over the past century. They often go by several other names, including *blocking models, variance component models, nested and split-plot designs, hierarchical linear models, multilevel models, empirical Bayes, repeated measures, covariance structure models,* and *random coefficient models*. They also overlap with *longitudinal, time series,* and *spatial smoothing* models. Mixed models are one of the most powerful and practical ways to analyze experimental data, and if you are a scientist or engineer, investing time to become skilled with them is well worth the effort. They can readily become the most handy method in your analytical toolbox and provide a foundational framework for understanding statistical modeling in general.

This book builds on the strong tradition of mixed model software offered by SAS Institute, beginning with PROC VARCOMP and PROC TSCSREG in the 1970s, to PROC MIXED, PROC PHREG, PROC NLMIXED, and PROC PANEL in the 1990s, PROC GLIMMIX in the 2000s, and more recently PROC HPMIXED, PROC LMIXED, PROC MCMC, PROC BGLIMM, and related Cloud Analytic Service actions in SAS Viya. We borrow extensively from *SAS for Mixed Models* by Littell et al. (2006) and Stroup et al. (2018). Mixed model software in various forms has evolved extensively and somewhat independently over the past several decades in other packages including R (lme4, lmer, nlme), SPSS Mixed, Stata xtmixed, HLM, MLwiN, GenStat, ASREML, MIXOR, WinBUGS/OpenBUGS, Stan, Edward, Tensorflow Probability, PyMC, and Pyro (web search each for details). The existence and popularity of all of these also speaks to the power and usefulness of mixed model methodology. Some differences in syntax, terminology, and philosophy naturally occur between the various implementations, and we hope the explanations and coverage in this book are clear enough to enable translation to other software should the need arise.

Mixed model functionality has been available in JMP since 2000 (JMP 4), and a dedicated mixed model personality in Fit Model was released in 2013 (JMP Pro 11). It continues to be an area of active development. The unique and powerful point-and-click interface of JMP, designed intrinsically around dynamic interaction between graphics and statistics, makes it an ideal environment within which to fit and explore mixed models. Analyzing mixed models in JMP offers some natural conveniences over any approach that requires you to write code, especially with regards to the engaging interplay between numerical and pictorial results of statistical modeling. To get an initial idea of how it works, let's dive right into our first mixed model analysis in JMP.

1.2 Cell Viability Example

Consider the cell viability data shown the previous section and contained in `Cell Viability.jmp`.

Using JMP

With the `Cell Viability` table open and active, from the top menu bar click *Analyze > Fit Model* to bring up a dialog box. On the left side, choose `Y`, then assign it to the *Y* role. Make sure the *Standard Least Squares* personality is selected in the upper right corner. Then select Chemical and Plate, and click *Add* to assign them to the *Construct Model Effects* box. In that box, select Plate, then click the red triangle ▾ beside Attributes and select *Random Effect*. You will see "& Random" added beside Plate in the box, confirming designation as a random effect. Click *Run* to fit the model.

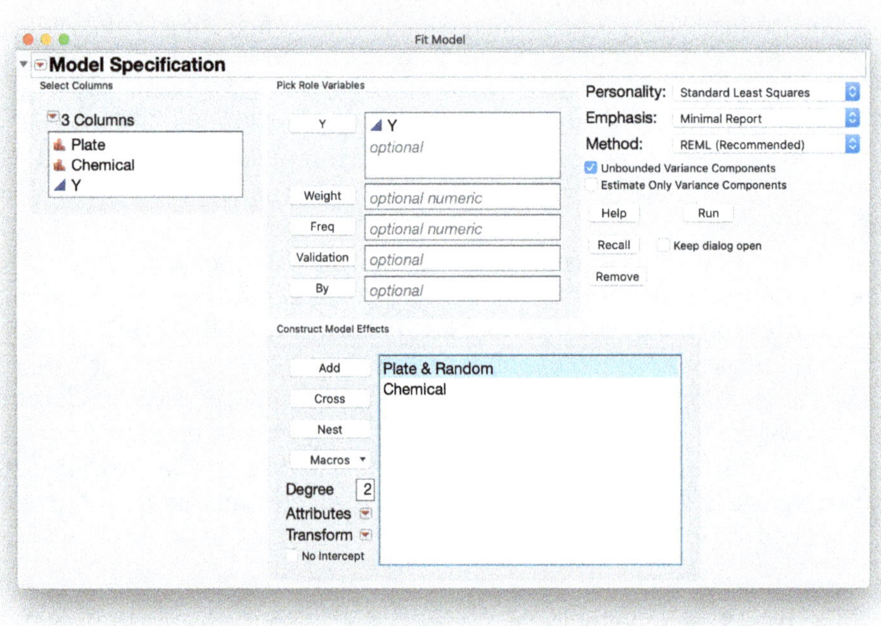

The model fitting results in Figure 1.2 are comprehensive with numerous statistics and details. We only focus on a few of the most important ones here and explore others in more depth in later chapters.

Figure 1.2: Mixed Model Results for Cell Viability Data

Cell Viability - Fit Least Squares

Response Y

Effect Summary

Summary of Fit

RSquare	0.866
RSquare Adj	0.854833
Root Mean Square Error	1.125126
Mean of Response	5.157143
Observations (or Sum Wgts)	14

Parameter Estimates

Term	Estimate	Std Error	DFDen	t Ratio	Prob>|t|
Intercept	4.9687502	0.65603	6.877	7.57	0.0001*
Chemical[A]	-0.943647	0.362424	4.104	-2.60	0.0583

Random Effect Predictions

REML Variance Component Estimates

Random Effect	Var Ratio	Var Component	Std Error	95% Lower	95% Upper	Wald p-Value	Pct of Total
Plate	2.3186713	2.9352253	2.2582795	-1.490921	7.3613718	0.1937	69.867
Residual		1.2659083	0.9881616	0.4210348	14.717247		30.133
Total		4.2011336	2.0225058	1.9623186	14.477028		100.000

-2 LogLikelihood = 53.837363129

Note: Total is the sum of the positive variance components.

Total including negative estimates = 4.2011336

Covariance Matrix of Variance Component Estimates

Iterations

Fixed Effect Tests

Source	Nparm	DF	DFDen	F Ratio	Prob > F
Chemical	1	1	4.104	6.7793	0.0583

Effect Details

In the **Parameter Estimates** box in Figure 1.2, the row beginning with "Chemical[A]" contains the estimate of the effect of Chemical A (-0.94) along with its estimated standard error (0.36). Note this standard error is computed accounting for the random effect Plate in the model. Taking the ratio of these two numbers produces a t-statistic (signal-to-noise ratio) of -2.60. The associated p-value is 0.058, just above the classical 0.05 rule of thumb for statistical significance. As emphasized in recent commentary (see *The American Statistician* (2019)), such a borderline "non-significant" result should be interpreted in conjunction with the effect estimate itself and how it relates to estimated levels of variability in the context of the experiment.

Not shown is the estimate of Chemical B, which is automatically set equal to the negative of Chemical A in order to identify the model using the traditional sum-to-zero parameterization for linear models. The statistics for Chemical B are therefore identical to those from Chemical A, but the main effect and t-statistic have opposite signs. Our main conclusion is that Chemical B is estimated to have an overall effect around 1.9 units higher than Chemical A.

The **REML Variance Component Estimates** box provides estimates of the variance components along with associated statistics. Here we see that the estimate of plate-to-plate variability is 2.3 times larger than within-plate (residual error) variability. Such a result speaks to the two primary sources of random variability in this experiment and prompts questions as to why plates are varying to this degree.

Key Terminology

The acronym REML refers to restricted (or residual) maximum likelihood, the best-known method for fitting mixed models assuming that any missing data are missing at random, and equivalent to full information maximum likelihood from econometrics. Refer to Stroup et al. (2018) for details and theory behind REML in mixed models.

All of the mixed model results help to answer various aspects of research questions involving these chemicals and the assay used to assess them. Note you can also obtain confidence intervals by clicking the small red triangle near the upper left corner of the report (just to the left of **Response Y**) then selecting *Regression Reports > Show All Confidence Intervals*. The 95% interval for the estimate of the Chemical A effect in this case is (-1.94, 0.05), just barely containing zero.

The red triangle menu is loaded with several additional analyses, including many graphical displays. A key philosophy behind the design of JMP is to utilize relevant interactive graphics directly alongside statistics. As one good example, click > *Row Diagnostics > Plot Actual by Predicted* to produce Figure 1.3.

Here the predicted values are based only on the fixed chemical effect from the model and not the random plate effect, explaining why there are only two distinct values on the X axis. A key aspect revealed by this plot is the increased variability in predictions corresponding to Chemical B on the right, as compared to those from Chemical A on the left. This is driven by the two lowest predicted values on the right. Selecting these two points in the graph and looking back at the table reveals they come from the first two plates. This is a reason to recheck that nothing unusual occurred on these two plates.

Figure 1.3: Actual by Predicted Values for Cell Viability Data

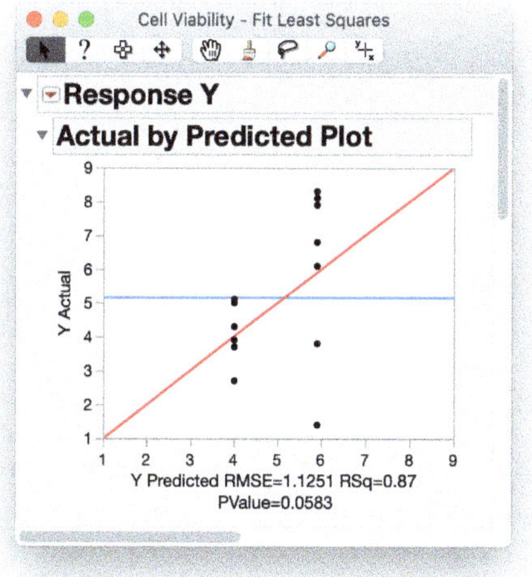

To track this further, let's plot the raw data. With the Cell Viability table in focus, click *Graph > Graph Builder*, assign Y to the *Y* axis drop zone, Chemical to the *X* axis drop zone, and Plate in the *Overlay* role. Then select the Line plot element and click *Done* to produce Figure 1.4.

The points in Figure 1.4 have similar orientation to the model-based ones in Figure 1.3, but now points occurring on the same plate are connected with a line. The four singleton points correspond to the four plates that contain only one observation, with the other one missing. The point in the far bottom right might be considered an outlier as well as influential in the estimate of plate-to-plate variability. A decision to potentially remove it depends critically on the quality of experiment protocol, and care must be taken to maintain data and research integrity.

The preceding analysis flow illustrates how to perform a basic mixed model analysis within JMP and the ease with which you can effectively utilize graphical displays to reveal potentially hidden or unusual patterns in your data suggested by mixed modeling. The combination of advanced statistical models and targeted graphics is a powerful one. In many cases you might want to do some graphical explorations in JMP before mixed modeling, and that is a great way to proceed as well.

Note it is also possible to analyze these data using *Analyze > Specialized Modeling > Matched Pairs*, which performs a classic paired *t*-test along with a rotated graph. However the four rows with missing values of Y are dropped and results are less efficient than those shown here. As one illustration of the difference, the degrees of freedom used to com-

Figure 1.4: Cell Viability Raw Data Plot

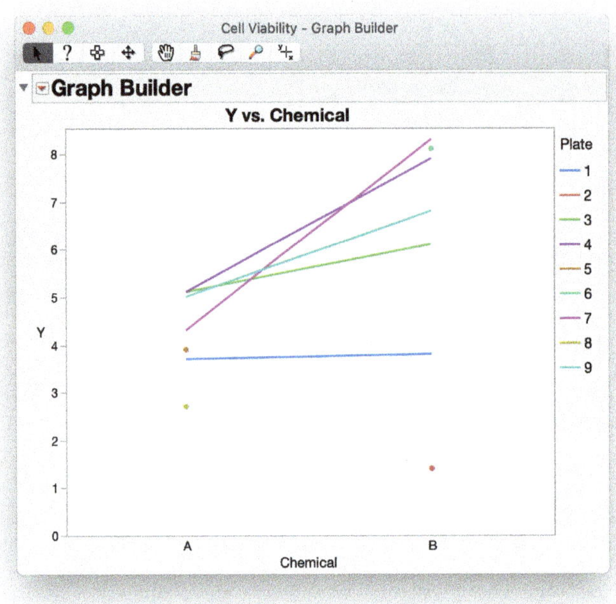

pute the p-values in the mixed model analysis are fractional (for example, 4.1 in the F test near the bottom of the output above). These are obtained using an advanced algorithm (Kenward and Roger, 1997) to more accurately approximate the small sample distributions of the statistics given the imbalance in the data due to the four missing values. A mixed model is able to handle missing data like this and deliver better results than a classical paired-t analysis.

In this example, the observed data that we analyze are the three columns in the Cell Viability table: the assigned levels of Chemical and Plate, and the response Y. All unknown parameters from the mixed model are estimated with REML using these quantities as inputs. Under key assumptions the estimated parameters enable us to make direct quantitative assessments of the causal effect of Chemical on Y amongst plate-to-plate and residual variability. Let's explore these assumptions in detail.

1.3 Mixed Model Assumptions

Several key assumptions are behind the validity of the preceding modeling results for the cell viability data. Continuing with this example as a prototype, we now describe the key statistical and structural form of these assumptions. We begin with a statistical description of a basic mixed model.

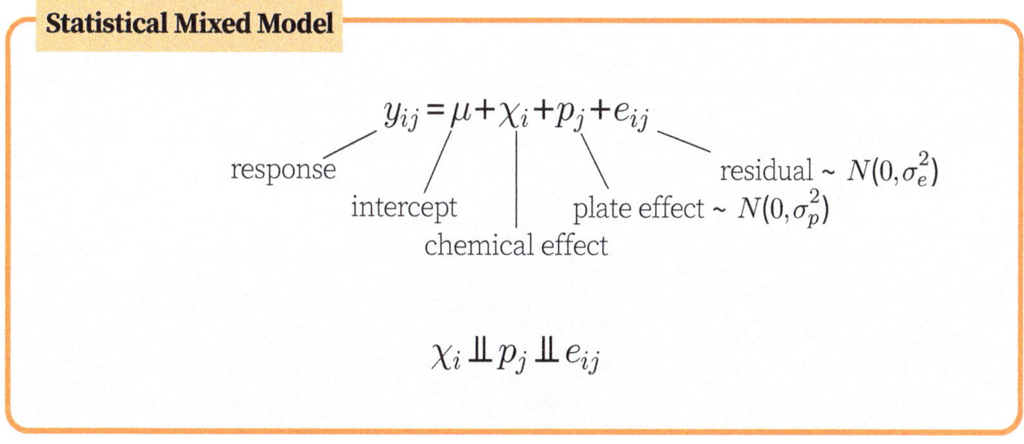

This is the simplest possible mixed model, with one fixed effect (Chemical) and one random effect (Plate). It is a linear mixed model, because it is an additive function of all primary components. The subscripts i and j index the individual observations; here $i = 1,2$ and $j = 1,\ldots,9$, and y_{ij} is the response for the ith chemical on the jth plate.

Each term on the right hand side of the model contains unknown parameters that we estimate from the data. We adopt the convention here and throughout the book that Greek letters denote fixed effects and Roman letters denote random effects.

The first fixed effect is μ, which models the central tendency of the data, also known as an *intercept*. We expect its estimated value to be near the simple mean of Y. For the fixed effect Chemical, we specify two parameters, χ_1 and χ_2, to model the effects of Chemical A and B, respectively. These are our primary parameters of interest for this experiment.

The notation $p_j \sim N(0,\sigma_p^2)$ is a shorthand way of stating that the random plate effect consists of independent and identical realizations from a normal (Gaussian) probability distribution with mean 0 and variance σ_p^2. The errors e_{ij} have the same form of probability assumption and serve as a catch-all for the numerous, small, unobserved effects driving variability of Y within each plate, also known as residuals. The notation $\chi_i \perp\!\!\!\perp p_j \perp\!\!\!\perp e_{ij}$ denotes statistical independence (Dawid, 1979) among its three components. Even though χ_i are considered fixed unknown parameters, the independence here refers to the treatment assignment mechanism of the levels of χ_i to the half-plates.

This completes the formal set of assumptions that we make when viewing a mixed model as a *statistical* model, suitable for assessing associational relationships and for making predictions. We have defined the full conditional probability distribution of y_{ij} given all elements on the right hand side of the model.

Given our randomized experiment setting, we can readily move from association to causality and infer the causal effects of Chemical and Plate on Y. This entails viewing the model as *structural* and assuming each term on the right hand side is *exogenous*, that is, wholly and causally independent of other variables in the system. We can depict this with a directed acyclic graph (DAG) as follows.

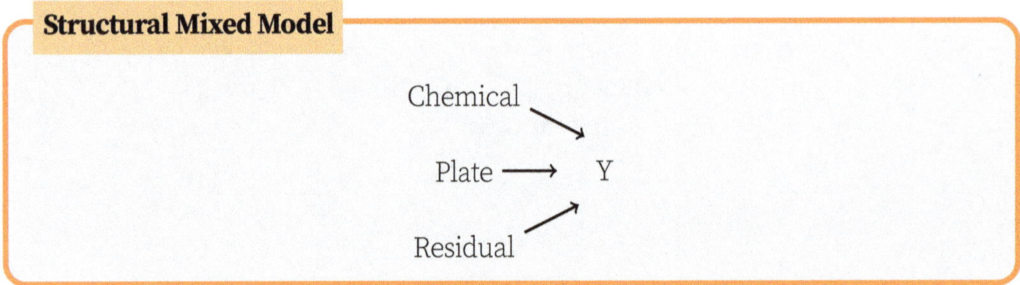

Note the direction of the causal arrows from the three causes to Y. Importantly, the absence of arrows into and between Chemical, Plate, and Residual indicates their exogeneity. Furthermore, the absence of any additional arrows into Y indicates there are no unmeasured causes or confounders besides those included in Residual. In addition, residual error is no longer just defined by algebraic subtraction, but consists of independent noise effects uniquely influencing each observed value of Y. Assumptions along these lines are required for causal inference. Refer to Pearl (2009), Heckman (2008), Imbens and Rubin (2015), Hernán and Robins (2020), and Chapter 10 for a comprehensive discussion.

It is critical that you fully understand the preceding modeling assumptions and their implications, keeping them in mind as you interpret modeling results. Strictly speaking, the assumptions may not be precisely true, but they do not need to be. As long as the assumptions provide a reasonably adequate approximation to the true data-generating mechanism, you can make sufficiently reliable associational and causal conclusions along with a statement of accompanying uncertainty.

For the cell viability example, the assumptions on p_j and e_{ij} made above imply that y_{ij} is normally distributed with a well-defined mean and covariance structure, and the validity of printed t-statistics are made under this assumption. This model typically would not be appropriate for a response that is nonnormal (e.g. binary, count, or time-to-event), but you can handle such situations with extensions such as the generalized linear mixed model discussed in Chapter 9, or with transformations of the data that enable better alignment with the underlying assumptions. You are also free to only adopt standard statistical assumptions or go further and make causal ones, depending on the objectives of your analysis.

As we proceed with various examples throughout the book, we will indicate various ways of checking the aforementioned assumptions. The methods can be statistical or graphical, and often involve analyzing deviations from fitted model predictions. Some assumptions, especially those for causal inference, are only indirectly or even fully untestable from observed data. For these you must rely on your scientific know-how and common sense, always maintaining a healthy degree of skepticism about how well your model approximates the true data-generating process of the system that you are studying.

As one example of the type of graph that is helpful for assumption checking, fit a mixed model on the **`Cell Viability`** data as in Section 1.2. Recall we left the previous analysis of these data wondering about a potential outlier. How well does our mixed model fit this value? Near the upper left corner of the analysis report, click the red triangle > *Save Columns* > *Conditional Pred Formula*. From the same menu also save *Conditional Residuals*. Return to the Cell Viability JMP table and note two new columns have been added. Click *Graph* > *Graph Builder*, assign **`Cond Residual Y`** to the Y axis and **`Cond Pred Formula Y`** to the X axis, to obtain a graph like Figure 1.5.

Figure 1.5: Cell Viability Conditional Residuals

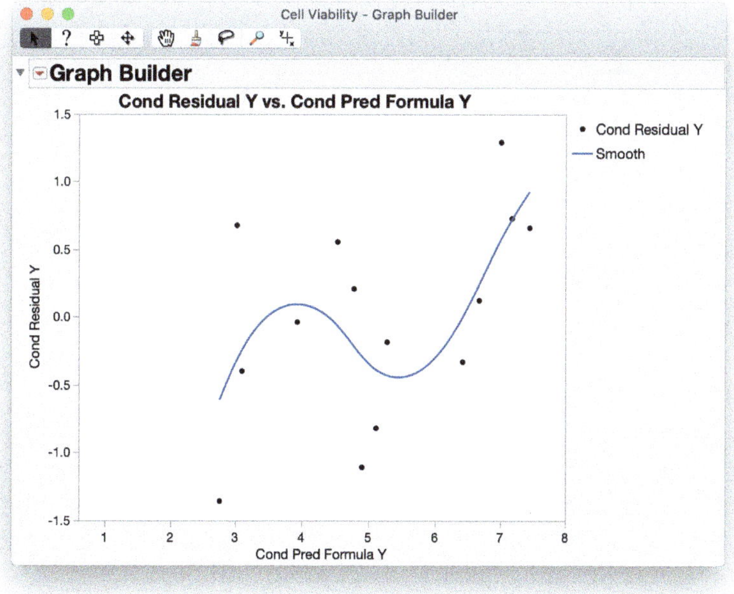

The residuals in Figure 1.5 are original Y values minus conditional predicted values (predictions including random effects). Graph Builder overlays a smooth curve by default. Note the range of the Y axis is from -1.5 to 1.5. This type of plot under usual mixed model assumptions should exhibit randomly scattered noise around a horizontal line at 0, with the fitted smooth curve also horizontal. Here the smooth curve is somewhat far from that ideal. If you bring up the previous raw data plot side-by-side with this one, you can interactively select points in one plot and see them highlighted in the other. The outlier apparent in the raw data graph corresponds to the lowest left point in this plot. The three largest residuals in the upper right correspond to the three largest Y values. The mixed model predictions are shrunken somewhat towards the overall mean. This is a small data set with a fair amount of noise, and given the lack of fit, any final conclusions should be considered tentative.

1.4 Nominal and Continuous Variables

In the Cell Viability example from the previous sections, the Chemical and Plate variables in the JMP table are assigned modeling type *Nominal* (having named, discrete, unordered levels), as indicated by the red histogram icon, whereas Y is *Continuous* (having numerical values with an implied distance measure between them), as indicated by the blue triangle icon. The Nominal type of Chemical and Plate is crucial while modeling, as it notifies JMP to create distinct levels when constructing the parameters to be estimated. This is particularly important for variables like Plate, whose values in the table are numeric.

When first entering a numeric variable in JMP, it is assigned by default to be Continuous (blue triangle). To change this attribute in any JMP table or dialog, click on the variable's icon and select the desired modeling type such as Nominal. A third possibility is Ordinal (green increasing histogram), but it is typically not needed for the common mixed models in this book.

Nominal or Continuous modeling type specification in JMP is in contrast to the use of a CLASS statement in various SAS/STAT procedures like PROC MIXED and PROC GLIMMIX. JMP has no CLASS statement; rather, you must prespecify effects to be Nominal or Continuous before specifying them in a modeling dialog. Note it is possible to create effectively identical models by converting a nominal variable to continuous ones with values 0 or 1, also known as indicator variables, dummy variables, or one-hot encoding.

Using independent variables that are continuous in a mixed model produces regression-style models. Such variables are often referred to as *covariates* or *regressors*. Standard regression analysis views such variables as fixed effects and the estimated parameters multiplying them in a linear model correspond to slopes. An effective mixed model extension of linear regression enables you to specify random slopes corresponding to meaningful clusters of the data, a type of model we refer to as *random coefficients*. These are a form of hierarchical linear models popular in social science and econometrics applications; see Chapter 5.

1.5 Experimental Units and Blocking, Cell Growth Example

When considering data from an experiment a fundamental question to ask is: On precisely what entities have treatment levels been applied or randomly assigned? In our cell viability example, at first glance you might consider the entities to be the plates or the individual cells growing on them; however, neither of these is exactly right with respect to the Y responses. Such considerations involve the fundamental concept of an *experimental unit*.

Key Terminology

Experimental Unit The smallest entity to which a treatment is independently assigned. In the cell viability example, the experimental unit for Chemical is a half-plate.

Note there is no variable in the cell viability table for half plate even though it is the experimental unit for Chemical. This is because half plates correspond to the rows in the table itself and JMP and other common mixed modeling software is able to recognize this. Now suppose you subdivide samples into three replicates and triple the number of measurements and rows in the table. You would then want to add new columns like HalfPlate and Replicate to the table to designate the experimental units for Chemical and the replicate numbers, respectively.

While our cell viability example is relatively simple, the data and experiments that you commonly analyze are likely more complex. Consider the data in Figure 1.6, which represent an extension of the cell viability experiment and are available in `Cell Growth.jmp`.

The Plate, Chemical, and Y data values in `Cell Growth` are identical to `Cell Viability`, but there are now two additional factors: Incubation and Batch. Our experiment objectives are extended to investigate the effect of three different incubation periods (short, medium, and long). In addition, the incubation chamber has room for only three plates, and the Batch variable indicates which plates are incubated together.

Given what we know so far about mixed models, how would you designate the new factors Incubation and Batch with regard to being fixed or random effects? The three levels of Incubation are ordered and constant for this experiment, and we want to directly compare how they change Y. Incubation is thus considered to be a fixed effect. Furthermore, notice that levels of Incubation are applied to entire plates, so the experimental unit for Incubation is a plate. We therefore now have two different sizes of experimental units: plates and half-plates. This arrangement is known as a *split-plot design*, which we cover in detail in Chapter 3.

What about Batch? We can naturally consider the effect of a particular run in the incubation chamber to be transient and sampled from a theoretical population of such runs. Batch is therefore a random effect. The fact that Batch groups sets of three plates brings us to another very important concept in experiment design.

Figure 1.6: Cell Growth Data

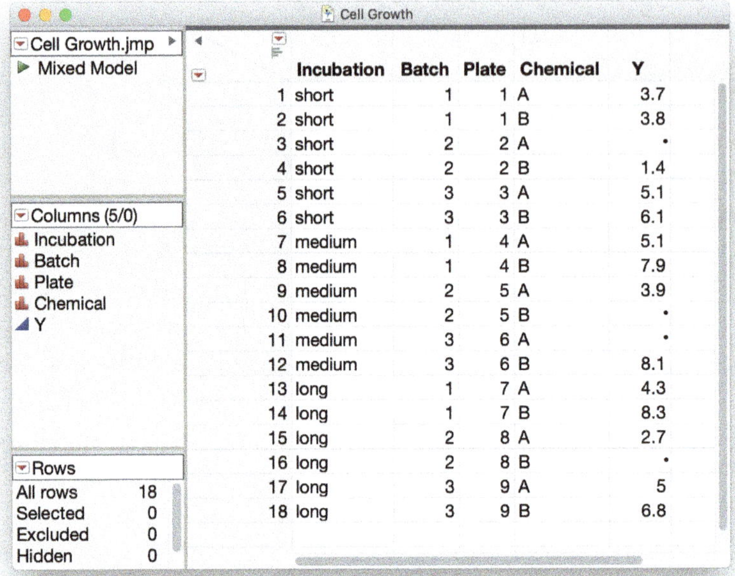

	Incubation	Batch	Plate	Chemical	Y
1	short	1	1	A	3.7
2	short	1	1	B	3.8
3	short	2	2	A	•
4	short	2	2	B	1.4
5	short	3	3	A	5.1
6	short	3	3	B	6.1
7	medium	1	4	A	5.1
8	medium	1	4	B	7.9
9	medium	2	5	A	3.9
10	medium	2	5	B	•
11	medium	3	6	A	•
12	medium	3	6	B	8.1
13	long	1	7	A	4.3
14	long	1	7	B	8.3
15	long	2	8	A	2.7
16	long	2	8	B	•
17	long	3	9	A	5
18	long	3	9	B	6.8

Key Terminology

Block A group of experimental units that are similar in some fashion, distinguishing them from other groups of experimental units. In the cell viability example, each level of Plate defines a block consisting of two half-plates. In the cell growth example, Batch defines blocks with three plates each.

When designing and analyzing your experiments, there are several reasons why you might want to create blocks or batches. Experimental units may naturally occur in clusters or groups. For example, in the cell growth data above, plates naturally group half-plates, and the incubation chamber size restriction requires us to use batches of three plates. Another type of blocking example is different shipments or lots of a raw material or reagent from a supplier. Blocking typically enables you to control for known sources of variability and obtain more precise inferences on the fixed effects.

A good question to ask is: What features of your experiment are the key sources of variability and covariability in the responses from one observation to the next? Identify these features and then model their effects by, for example, creating batches to consolidate the groups of similar runs. Now any differences between the batches will give you a measurement of the variability that comes from this source, and you can better estimate uncertainty around your estimates for the treatment effects once you've accounted for it. This is the purpose of a well-thought-out design—partition out the variability that can't be controlled (but can be explained), and then you will have more precision for the rest of your conclusions.

Blocking factors can have many different names. Blocks could be called *lots* or *batches*, or they could be called by the effect they are modeling, such as person or school. *Strata* is another popular term. The point is simply that you are grouping the experimental units so that they are more similar within groups and less similar (often due to effects that you cannot control) between groups.

Occasionally blocks can account for significant dissimilarity. A classic example is a litter of pups from a common dam who are competing for a fixed amount of resources (food or familial attention). In this type of situation, we might expect the measurements within a block to be negatively correlated, and in the mixed model analysis, the estimated variance component would be negative.

Blocking can often be very effective when applied in two different directions. Common examples include rows and columns in a field plot or 96-well plate. Chapter 4 contains an example of a *row-column design* (Latin Square) implementing this concept. This type of blocking also relates to general considerations of *spatial variability*; see Chapter 7.

Another type of blocking can occur when you observe repeated measurements on the same experimental units, for example, in a longitudinal or time series study on individuals. Repeated measures are usually no longer independent of each other, and a mixed model is a great way to handle this source of correlation. New complexities arise, as there can be different levels of measurement and different types of covariance structures; see Chapters 5 and 6.

In general the goal of blocking is to control for specific sources of variability and thereby achieve more accurate and precise inferences. Blocking can often make considerable difference in modeling results. Because blocks are usually assumed to arise from a population of effects, you typically will want to declare all blocking factors as random effects in your mixed model analysis.

1.6 Confounding

Suppose the data for the cell growth data are the altered version in Figure 1.7.

The final three columns in Figure 1.7 are identical to the Cell Growth data in Figure 1.6, but the first two columns are different. Can you spot the problem?

Figure 1.7: Cell Growth Confounded Data

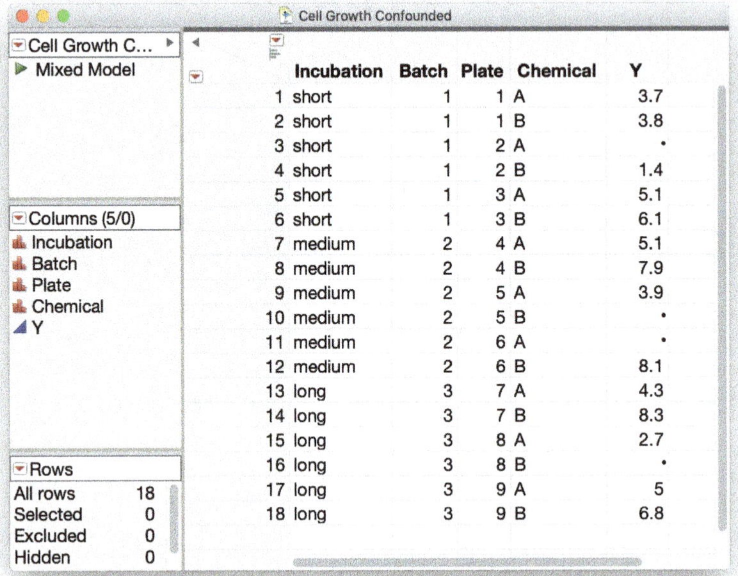

Cell Growth Confounded

Cell Growth C...
Mixed Model

Columns (5/0)
Incubation
Batch
Plate
Chemical
Y

Rows
All rows 18
Selected 0
Excluded 0
Hidden 0

	Incubation	Batch	Plate	Chemical	Y
1	short	1	1	A	3.7
2	short	1	1	B	3.8
3	short	1	2	A	•
4	short	1	2	B	1.4
5	short	1	3	A	5.1
6	short	1	3	B	6.1
7	medium	2	4	A	5.1
8	medium	2	4	B	7.9
9	medium	2	5	A	3.9
10	medium	2	5	B	•
11	medium	2	6	A	•
12	medium	2	6	B	8.1
13	long	3	7	A	4.3
14	long	3	7	B	8.3
15	long	3	8	A	2.7
16	long	3	8	B	•
17	long	3	9	A	5
18	long	3	9	B	6.8

Key Terminology

Confounding The indistinguishability of two or more effects, given a model and data set. Such effects are *confounders* of each other. In the example data above, Incubation and Batch are *complete* confounders. Confounding can also be *partial,* in which portions of effects are unable to be disentangled from portions of other effects given the model and data.

Confounding is a danger lurking in many statistical and structural models of data, and you should be constantly on the lookout for it. Sometimes it is innocuous, as in the case we have already encountered involving the parameters μ, χ_1, and χ_2 in the simple mixed model for the cell viability data. We resolve this by imposing the sum-to-zero constraint $\chi_1 = -\chi_2$. Similar constraints typically work for complex linear effects and their interactions, ensuring identifiability of the model.

The preceding example, however, is much more serious. The data provide no way to separate the effects due to Incubation and Batch. Worse would be a case where only one of the variables is observed, leading to likely incorrect conclusions about the true magnitude of that effect. The mistake here is in the experiment design, and a principal goal of good design is to avoid confounding like this. Note in the original Cell Growth data in Figure 1.6, Incubation and Batch are nicely *orthogonal*, with each incubation level occurring exactly once within each batch.

In many data sets, especially observational ones, partial confounding is unavoidable. When confounding happens, the best you can do is understand its precise nature, determine exactly how effects are aliased, and limit your conclusions appropriately. Refer to Hernán and Robins (2020) for helpful insights on confounding in the context of causal inference, including such difficulties as unmeasured confounders.

1.7 JMP and JMP Pro

A more advanced version of JMP is available as *JMP Pro*. The *Fit Model* platform in JMP Pro adds a dedicated *Mixed Model* personality, and specifying mixed models in it is a bit different from the *Standard Least Squares* personality we use above in the cell viability and growth examples. You can optionally still use Standard Least Squares in JMP Pro. Certain mixed model functionality is only available in JMP Pro, including random coefficient models (Chapter 5), repeated measures models (Chapter 6), and spatial models (Chapter 7).

The following two boxes provide breakdowns of functionality available in the Standard Least Squares and Mixed Model personalities.

Using JMP

Once you run an analysis in JMP, you will find many drill-down options from the red triangle at the top left of the results report window. When using the *Standard Least Squares* personality, you will find the following options from the drill-down menu in the results report. For more information, go to `https://www.jmp.com/help` and search in the *Fitting Linear Models* section.

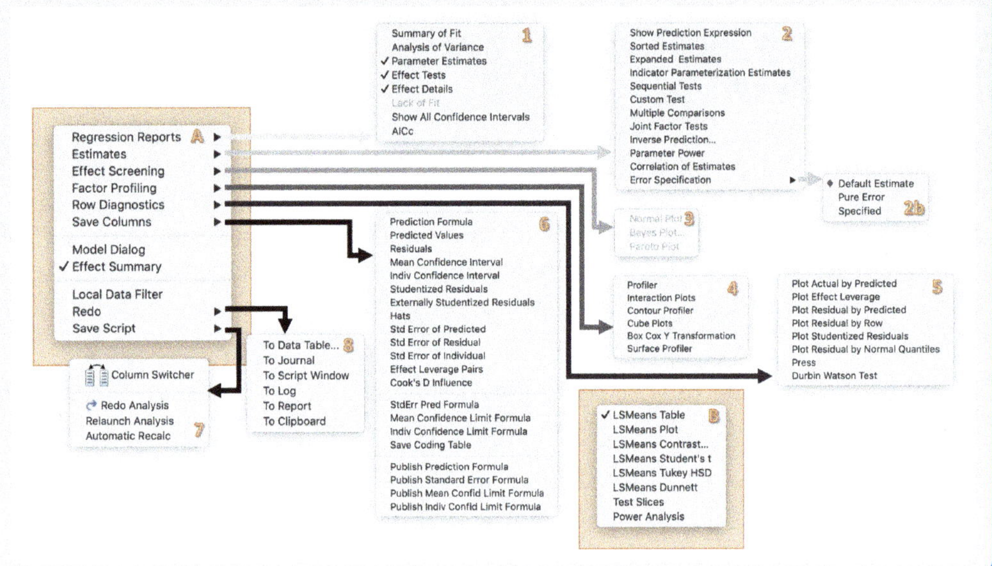

Once you run an analysis in JMP Pro, you will find many drill-down options from the red triangle at the top left of the results report window. When using the *Mixed Model* personality, you will find the following options from the drill-down menu in the results report. For more information, go to `https://www.jmp.com/help` and search in the *Mixed Models* section.

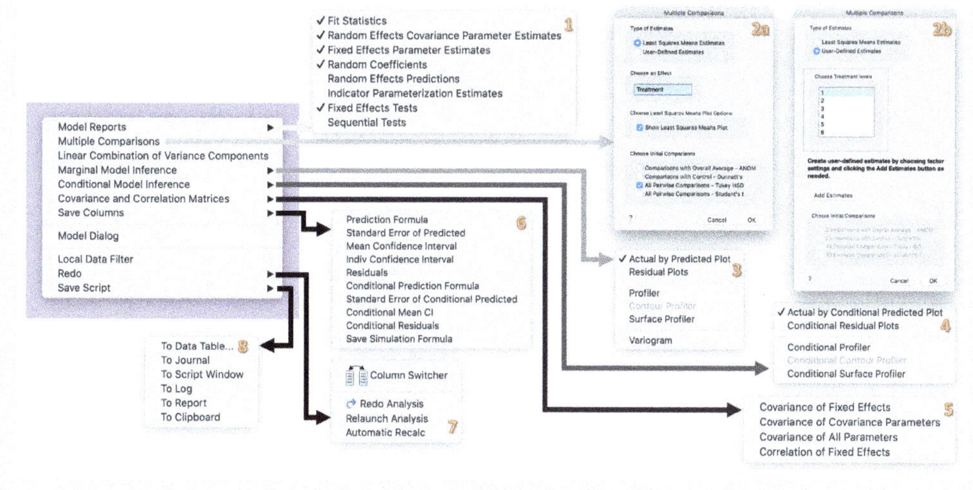

1.8 Exercises

1. In the output for the mixed model analysis of the cell viability data in Section 1.2, review each box of results and briefly describe the statistics displayed in each. Include question marks for the ones you do not yet understand.
2. In the cell viability data from Section 1.2, exclude the largest outlier for Chemical B and rerun the analysis. What type of difference does this outlier removal make in the results? Under what conditions would such exclusion be justified?
3. In the cell viability data, exclude or delete the rows with missing Y values and rerun the analysis. How do results change? Now analyze the data with *Analyze > Specialized Modeling > Matched Pairs*. How do the results from this analysis compare the previous one?
4. Fit an appropriate mixed model to the cell growth data from Section 1.5 and interpret key results.
5. Fit a mixed model to the confounded cell growth data from Section 1.6 and compare results with the previous exercise. What is the effect of confounding?

Chapter 2

ANOVA with a Single Blocking Effect

In this chapter, we closely examine experiments with one random blocking effect, like the Cell Viability example of Chapter 1. We begin by highlighting a few example scenarios in Section 2.1, then we introduce new statistical concepts and the statistical models needed in this chapter in Section 2.2. Finally, we work through the introduced examples, first with instructions for running these models in JMP and JMP Pro and then exploring and interpreting the output.

2.1 Motivating Examples

We explore the following two examples.

Metal Bond Breaking — A fabrication company uses various metals (the treatment) as bonding agents to bond pieces of another composition material together. The company is interested in studying the pressure required to break the bonds. The experiment is conducted by taking ingots of the composition materials (seven in total) and subdividing them so that each treatment can be randomly assigned to a piece from each ingot.

Balanced Incomplete Blocks — A researcher has four blocks of material, and each block can be used for three runs. The researcher has four treatment levels to test. The treatments are randomized to the blocks so that each treatment shows up in three of the four blocks, subject to the constraint that each pair of treatments will be used in at least one block together.

2.2 Blocking Designs and Skeleton ANOVA

What makes models with random effects so special? As described in Chapter 1, by treating a factor as a random effect, we are able to explain a known source of variation in the response and possibly improve the power of the primary tests of interest. A very common type of random effect is a block, which groups experimental units in some fashion. By considering blocks random we can make treatment comparisons applicable over the entire theoretical population of blocks, rather than making inference just for those observed.

Consider our first example, using three metals as bonding agents and testing which metal creates the strongest bonds. The company wants to determine differences in the bond strength of nickel, iron, and copper as bonding agents. We could design this experiment to measure bond strength on several independent samples of each metal bonding agent application. If we choose seven experimental units for each of the three metals, we have 21 samples in total.

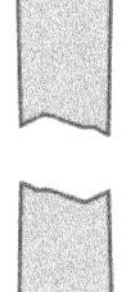

Prepare the experimental unit and assign the Treatment Level (i.e., which metal?)

Metal = Copper

Conduct the experiment: test the breaking strength of the metal bond.

If each experimental unit truly is selected independently of the others (none of the samples have anything extra in common, like coming from a certain batch or being chosen based on a certain order) and these experimental units are then randomly assigned to receive one of the metal bonding treatments (again, completely randomly), then this design is called a *Completely Randomized Design (CRD)*. We have one fixed treatment factor (metal) and seven replicates per treatment. The term *replicate* indicates the sample size within one treatment group or treatment combination. We generally use *sample size* to indicate the total sample size of the entire experiment.

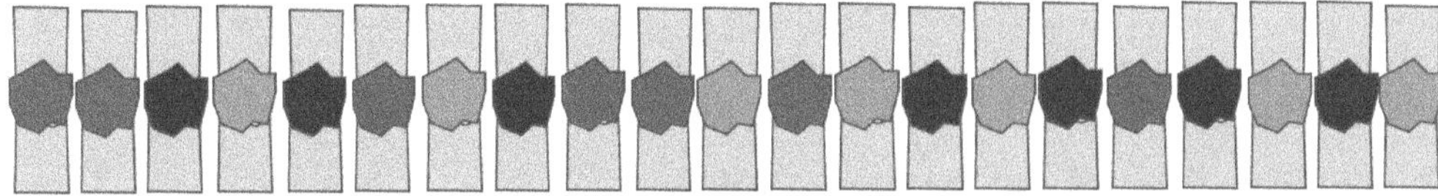

Balanced CRD and One-Way ANOVA

For any CRD with one fixed treatment factor and no random effects, we can write the statistical model as follows.

Basic One-Way ANOVA Model

$$y_{ij} = \mu + \alpha_i + e_{ij}$$

y_{ij} is the continuous response variable.
μ is the intercept.
α_i is the effect of the ith level of the treatment.
e_{ij} is the residual error and $e_{ij} \sim N(0,\sigma^2)$.

This is the classic one-way analysis of variance (ANOVA) model popularized by Fisher in 1925 (Fisher, 1925). The term *one-way* refers to a single classification scheme defined by the levels of α. The term *analysis* means to examine in detail the elements or structure of something, and Fisher's brilliant method is to explain variability in an observed response using levels of a known factor. This factor is a single fixed effect in our mixed-model terminology.

When each treatment group has the same number of replicates, we refer to the design as *balanced*. If that were not the case (due to resource limitations, lost measurements, or missing values in the data), we use the term *unbalanced*. We still write the corresponding one-way ANOVA model in the same way. The only difference is in the limits of the subscripts defining the replicates within groups.

Blocking Designs

We now extend the CRD and accompanying ANOVA model to include a blocking effect, as in the initial Cell Viability example of Chapter 1. In our bond example, the material on which we are making the metal bonds is made available in rectangular bars (called ingots). We believe that the material composition making up a single bar is very consistent, so any pieces that we remove from the ingot will be similar to each other. However, the chemical composition of the material varies somewhat from ingot to ingot. For this reason, we would like to compare the metal bonding agent treatments within each ingot. Said another way, we want to compare the metal bonding agent treatments after explaining away any differences that might be coming from differences in the ingots themselves. Because we have three treatment groups (the three metals), we cut three pieces from each ingot.

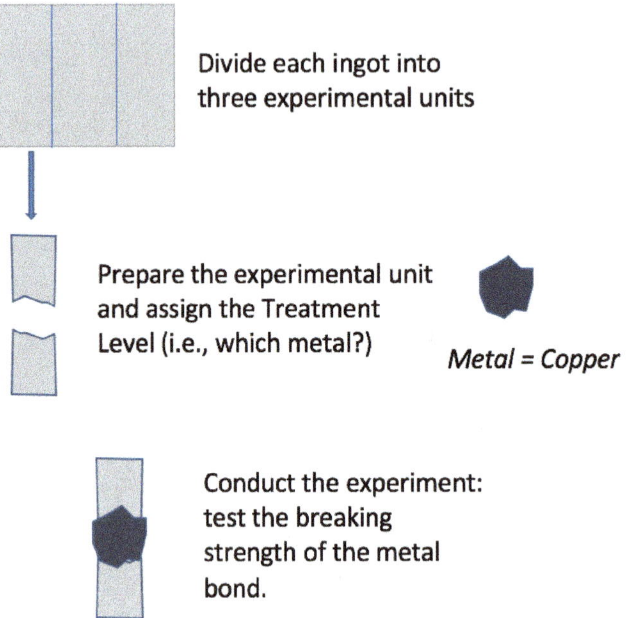

Each piece is now an experimental unit for each treatment, and the pieces are blocked by the ingot.

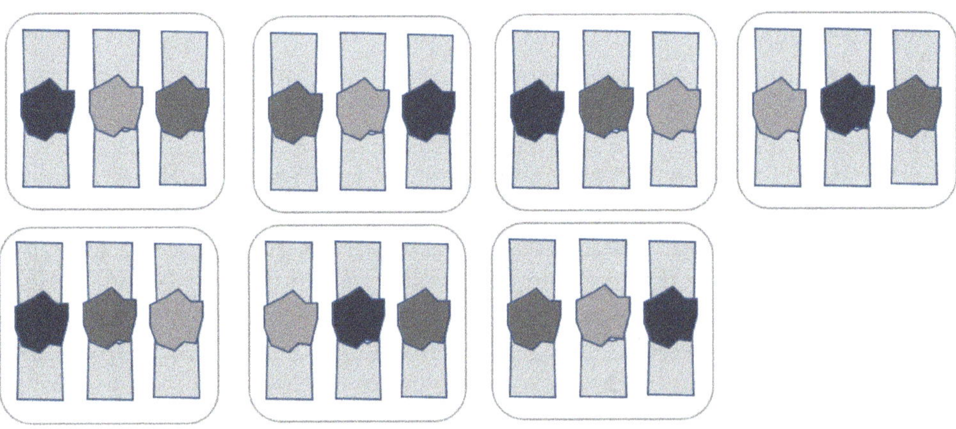

We use this term *blocked* to indicate that the experimental units are in groups defined by some characteristic that we expect to have similarity within the group and dissimilarity between groups.

Each metal occurs in a random order in each block. The random assignment is an assumption of the model and is important. If the order was not random, then there might be some other feature about the order of the treatment assignment which we are ignoring. If, for example, we thought there might be an effect of the edges versus middle of the ingot on the breaking strength for those bonds, then we might also include the Position as another blocking variable in our model. It is very important, when determining the effects to include in your experiment, to consider all sources of non-randomness or variability. If you can plan for these, you will increase your ability to gain insights from the data after you have run the experiment and avoid confounding.

Another point to notice is that each treatment occurs in each block. This, along with proper randomization, defines a *Randomized Complete Block Design* (RCBD). If you have limited resources and cannot run an RCBD or if data are deleted or you are not able to retain all of the planned runs of the experiments, you can still run an effective experiment. The *Balanced Incomplete Block Design* (BIBD) is a reduced version of an RCBD that allows the blocks to contain fewer experimental units than there are treatment levels. For example, it might be feasible to split an ingot into only two pieces for use with our three metal treatments.

The one additional requirement for a BIBD is that each pair of treatments must show up in the same number of blocks together. In our case, if nickel and copper are in three blocks together, then nickel and iron need to be in three blocks together and copper and iron need to be in three blocks together. There are a limited number of combinations of numbers of blocks and numbers of treatments for which there is a corresponding BIBD. The benefit of a BIBD, when an RCBD is not possible, is that the treatment effects can be completely separated from the effect of the blocks. This feature is called *orthogonality*, and it is the opposite of confounding. Having the treatment comparisons completely orthogonal to the block differences means that we will not confound these effects with each other.

In the case where both an RCBD and a BIBD are not possible, there is one more planned design that attempts to retain the least confounding possible: the *Partially Balanced Incomplete Block Design* (PBIBD). Like a BIBD, this design requires the treatment levels to show up in blocks together, but in a PBIBD some treatment levels will be in, say, 5 blocks together, while the rest of the treatment levels are in, say, 4 blocks together. This is an additional level of unbalance, but the design is still maximizing the amount of orthogonality possible under a limited number of runs and can estimate treatment effects without confounding.

For an RCBD, BIBD, PBIBD, or any general experiment with one fixed treatment factor, applied at random across one random blocking factor, the statistical model can be written as follows. This model applies to balanced and unbalanced designs and to designs with missing data, but it will not apply to data that do not meet the stated assumptions (like count data or other non-normally distributed responses).

One-Way ANOVA Model with One Random Blocking Effect

$$y_{ij} = \mu + \alpha_i + r_j + e_{ij}$$

y_{ij} is the continuous response variable.
μ is the intercept.
α_i is the effect of the ith level of the treatment.
r_j is the effect of the jth level of the random (blocking) factor and $r_j \sim N(0,\sigma_r^2)$.
e_{ij} is the residual error and $e_{ij} \sim N(0,\sigma^2)$.

This is the same mixed model we used in Chapter 1 for the Cell Viability example, but now applied to the various blocking designs.

Skeleton ANOVA

A helpful practice when designing mixed models is to write down your understanding of the blocking features and treatment design in a design table. This was first formalized by Stroup (2012), in a section humorously named "What Would Fisher Do?". The idea is based on a comment by Fisher on the *Complex Experiments* presentation by Yates (1935). Following Stroup (2012), as well as Federer (1955) and Milliken and Johnson (2009), we refer to the two distinct parts of the data structure as the *experiment design* and the *treatment design*, and we write these design elements in a table in order to clearly lay out the model terms that our experiment and treatment designs dictate.

This table is a trimmed down version of the classic ANOVA table, which typically involves a complete listing of sums of squares, mean square, their expected values, and F-statistics. We instead focus just on *degrees of freedom*, the number of experimental units that are free to vary for that particular effect. As such, we refer to it as a *Skeleton ANOVA*.

We begin by considering all experimental unit and blocking factors and list these in a column entitled *Experiment Design*. For the Cell Viability example introduced at the beginning of Chapter 1, we have Plate as a block and Half-Plate as the experimental unit for Chemical. The degrees of freedom (*df*) for effects like Plate are calculated by taking the number of levels of that source and subtracting 1, assuming none of the data are missing. Subtracting 1 is done to account for an intercept in the model. Here we have 9-1=8 degrees of freedom for Plate. When one source is nested within another, take the number of levels from factor inside of the nesting and subtract 1 from that before multiplying times the number of outer nested groups. In this case we have (2 half-plates - 1)*(9 plates) to arrive at 9 *df* for Half-Plate. Compute the total *df*, which should equal the full planned sample size minus 1.

To create the Skeleton ANOVA, insert treatment effects above the lines in the Experiment Design that describe their experimental units. Here we insert Chemical above Half-Plate, assign it 2-1 = 1 degree of freedom, and then subtract this *df* from Half-Plate. Half-Plate then becomes the Residual error since it is the smallest size experimental unit. The final table is as follows:

Experiment Design		**Skeleton ANOVA**	
Source	***df***	**Source**	***df***
Plate	9-1=8	Plate	8
		Chemical	2-1=1
Half-Plate	(2-1)*9=9	Residual	9-1=8
Total	18-1=17	Total	17

Using this Skeleton ANOVA table, we see all of the terms that we need to include in our model and the degrees of freedom for each. Terms in the Experiment Design are random effects, and the rest are fixed. Note the Residual error is technically a random effect but usually does not need to be specified when setting up a mixed model in software because it is automatically included. Writing down the Experiment Design and Skeleton ANOVA can be a big help in ensuring that you've considered the important blocking and treatment elements of your experiment and are ready to correctly specify your model. It is also a great thing to share with collaborators to ensure that everyone understands how the mixed model is set up.

2.3 Metal Bond Breaking Example

Recall our metal bond-breaking experiment, described in section 2.2. These data were originally published in Wackerly et al. (1996) and are in a file called `bond.jmp`. The data table is shown in Figure 2.1, along with bar charts of the `ingot` and `metal` factors and a histogram of the `pressure` response. The data corresponding to the iron metal are highlighted in the distribution output and also in the data table shown in Figure 2.1. In JMP, we can visually explore the impact of the metal bonding agent on the pressure required to break the bond by clicking on the bars corresponding to the other metals.

This design is a Randomized Complete Block Design (RCBD). Here we block by `ingot`, and we randomly assigned the three levels of the treatment `metal` to pieces obtained by breaking the ingots into thirds.

The Skeleton ANOVA is structured just like that for the Cell Viability example, and the mixed model is also the same.

Figure 2.1: Bond Data Table and Variable Distributions

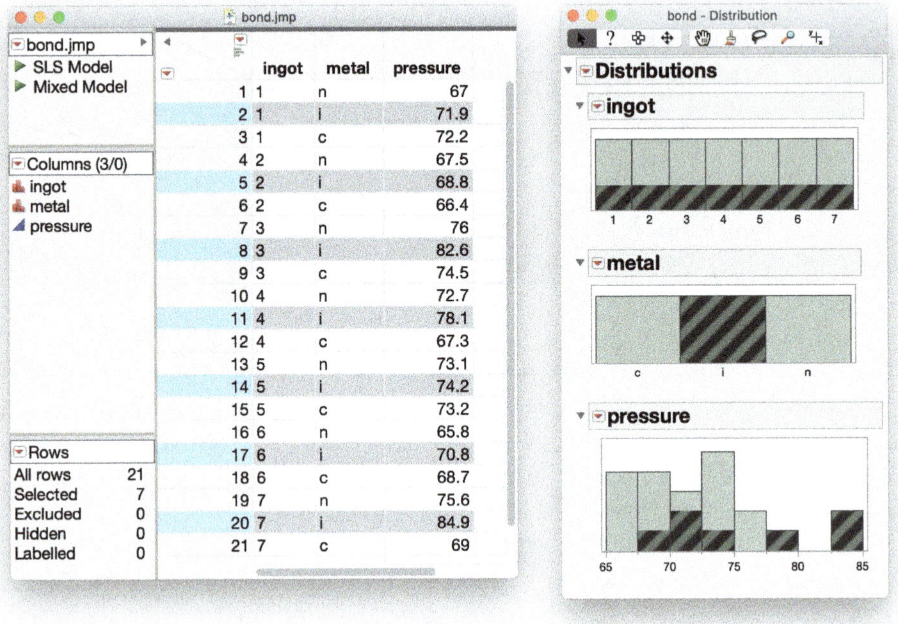

	ingot	metal	pressure
1	1	n	67
2	1	i	71.9
3	1	c	72.2
4	2	n	67.5
5	2	i	68.8
6	2	c	66.4
7	3	n	76
8	3	i	82.6
9	3	c	74.5
10	4	n	72.7
11	4	i	78.1
12	4	c	67.3
13	5	n	73.1
14	5	i	74.2
15	5	c	73.2
16	6	n	65.8
17	6	i	70.8
18	6	c	68.7
19	7	n	75.6
20	7	i	84.9
21	7	c	69

Experiment Design		**Skeleton ANOVA**	
Source	***df***	**Source**	***df***
Ingot	7-1=6	Ingot	6
		Metal	3-1=2
Ingot-Third	(3-1)*7=14	Residual	14-2=12
Total	21-1=20	Total	20

RCBD Statistical Model for Metal Bond Breaking

The statistical model for this example is

$$y_{ij} = \mu + \alpha_i + r_j + e_{ij}$$

y_{ij} is the pressure required to break the bond of the ith metal in the jth ingot.
α_i is the effect of the ith level of the metal. α_1 is the copper effect, α_2 is the iron effect, and α_3 is the nickel effect.
r_j is the effect of the jth random ingot and $r_j \sim N(0, \sigma_r^2)$.
e_{ij} is the residual error and $e_{ij} \sim N(0, \sigma^2)$.

To specify and fit the mixed model in standard JMP, use the *Fit Model* platform with the *Standard Least Squares* personality and the *REML* method. In JMP Pro, you can follow this same path using *Standard Least Squares*, or you can use the more powerful *Mixed Model* personality. We provide instructions for both here.

JMP Instructions for Metal Bond Breaking Example

Use *Analyze > Fit Model* to define the model with one fixed factor and one random factor.

Enter **`pressure`** as the *Y* variable. Note that the default personality becomes *Standard Least Squares* because you've entered a continuous variable in the *Y* role. Keep this default.

Enter **`ingot`** and **`metal`** in the *Construct Model Effects* box. To do this quickly, you can use the shift key to select both **`metal`** and **`person`** from the list of Columns at the same time, and then click the *Add* button to add these to the *Construct Model Effects* box. By default, these are *Fixed Effects*. To make the **`ingot`** random, first select it in the *Construct Model Effects* box and then select the *Attributes* drop-down and select **`Random Effect`**. Now **`ingot`** has the "& Random" appearing to the right in the *Construct Model Effects* box.

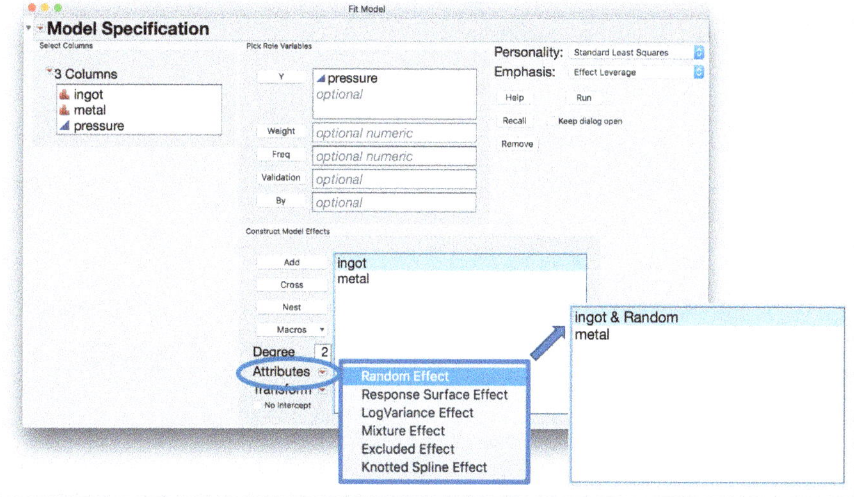

After clicking the *Run* button from the dialog box, the results will appear in a JMP Report window. We will describe some of this output here. For more information about the options and drill-down reports, see the JMP documentation (SAS Institute Inc. 2019a).

First we see information of the overall fit of the model in Figure 2.2. The ANOVA p-value associated with the fixed effect metal is p=0.0131, providing some evidence of significant

Figure 2.2: Summary of Fit and Variance Components for Metal Bond Breaking

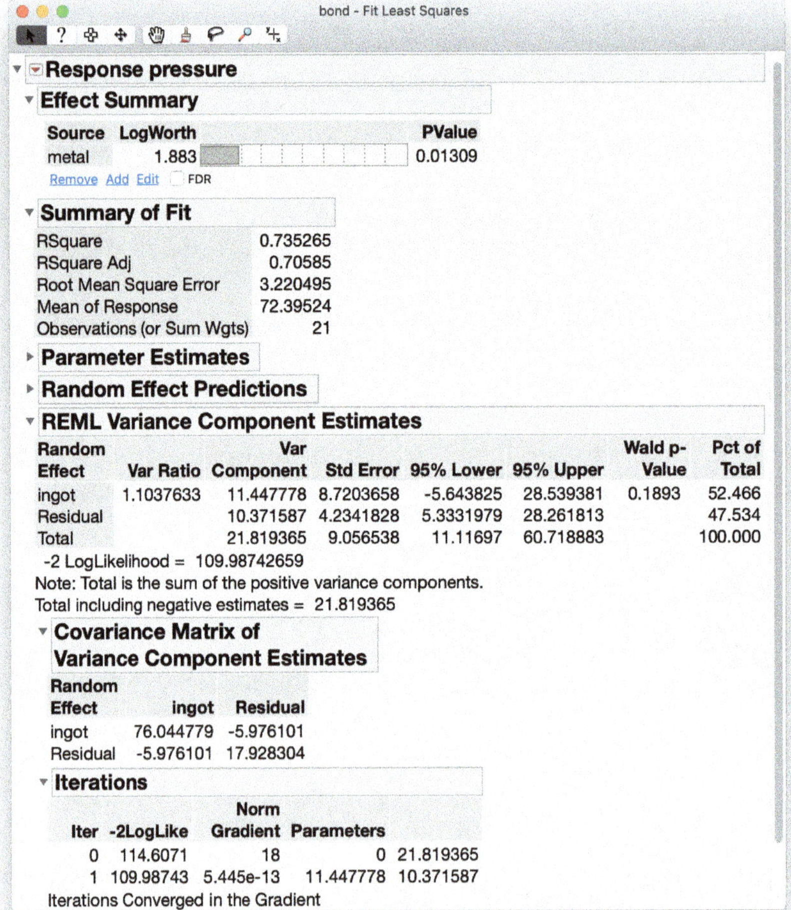

Source	LogWorth	PValue
metal	1.883	0.01309

RSquare	0.735265
RSquare Adj	0.70585
Root Mean Square Error	3.220495
Mean of Response	72.39524
Observations (or Sum Wgts)	21

Random Effect	Var Ratio	Var Component	Std Error	95% Lower	95% Upper	Wald p-Value	Pct of Total
ingot	1.1037633	11.447778	8.7203658	-5.643825	28.539381	0.1893	52.466
Residual		10.371587	4.2341828	5.3331979	28.261813		47.534
Total		21.819365	9.056538	11.11697	60.718883		100.000

-2 LogLikelihood = 109.98742659
Note: Total is the sum of the positive variance components.
Total including negative estimates = 21.819365

Random Effect	ingot	Residual
ingot	76.044779	-5.976101
Residual	-5.976101	17.928304

Iter	-2LogLike	Norm Gradient	Parameters	
0	114.6071	18	0	21.819365
1	109.98743	5.445e-13	11.447778	10.371587

Iterations Converged in the Gradient

differences between the metals. The corresponding *LogWorth*, which is $-log_{10}$ of the p-value, equals 1.8833. LogWorth is useful when comparing many small p-values to each other, since it is the exponent of standard scientific notation. In other words, LogWorth tells us how many zero places after the decimal point we need to go to get to the first significant digit in the p-value.

Looking further in Figure 2.2, we see from RSquare that 73.5% of the variation in the response variable, pressure, is explained by the model. We also see that the RMSE, or the square root of the mean square from the residual variance component, is 3.22 and that the mean of pressure is 72.4. Finally, we can confirm that a total of 21 observations are used in the analysis.

From the Variance Components section in Figure 2.2, we see that the variance explained by the ingot blocking variable is 11.4, or 52% of the total variation in the response variable, pressure. That leaves a variance of 10.4 for the residual (error), making up the remaining 48% of the total variation. The ratio of the ingot variance to the residual variance is therefore 1.1 — just slightly more variance is explained by the blocks than by the residuals within the blocks. The iteration history assures us that the REML estimation converged successfully.

Next, in Figure 2.3, we see the parameter estimates and, further down, a summary of the effect details for both the fixed and the random effects. Click on the gray triangle (disclosure icon) to the left of the name of the fixed effect, *metal*, under *Effect Details* to expand that section. The parameter estimates are the values that correspond to our statistical model.

Least Squares Means (LSMeans) provide a direct assessment of the three metal effects. LSMeans are not necessarily the actual group means as calculated by taking a mean of the data for that group. They can differ from the actual means under lack of balance, missingness of some data, and when there are other continuous effects in the model. LSMeans are typically more fair comparisons of groups than the true means are, however, because we are fixing them at specific values of the other effects and factors in the model, as if the model were balanced. For this reason, they are often called *adjusted means* or *population marginal means*. For decades, LSMeans have been the typical statistic reported for comparisons in scientific journals and textbooks for this type of model (a Standard Least Squares model estimated using REML). Refer to the JMP documentation for more information (SAS Institute Inc. 2019a). In addition, LSMeans are prominent in SAS/STAT mixed modeling procedures such as PROC MIXED and PROC GLIMMIX, and their documentation is also a good place to learn more about them.

In the results report, notice that, in addition to printing a list of the LSMeans for each treatment group, we see standard errors for those means. Corresponding lower and upper 2-sided 95% confidence interval limits can be added by right-clicking on this results table and selecting *Columns* and then *Lower 95%* and *Upper 95%*. We can also see those means and individual confidence intervals plotted by selecting the *Least Squares Means Plot* from the red triangle for `metal` under *Effect Details*. Note that these individual confidence intervals hint at differences between the group means, but they do not correspond directly to the statistical tests of differences. To conduct a statistical test to see whether the model-predicted means differ by treatment group, choose any of the testing options from the red triangle next to the `metal` under the Effect Details section of the report. We have the option to specify specific contrasts or to compare all pair-

Figure 2.3: Parameter Estimates and LSMeans for Metal Bond Breaking

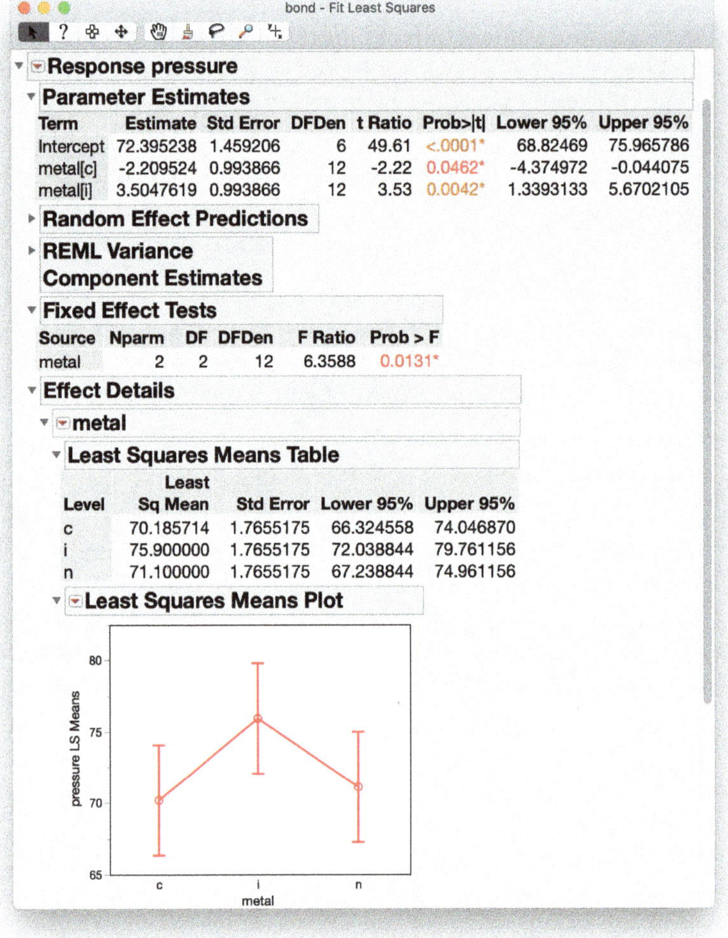

Term	Estimate	Std Error	DFDen	t Ratio	Prob>|t|	Lower 95%	Upper 95%
Intercept	72.395238	1.459206	6	49.61	<.0001*	68.82469	75.965786
metal[c]	-2.209524	0.993866	12	-2.22	0.0462*	-4.374972	-0.044075
metal[i]	3.5047619	0.993866	12	3.53	0.0042*	1.3393133	5.6702105

Source	Nparm	DF	DFDen	F Ratio	Prob > F
metal	2	2	12	6.3588	0.0131*

Level	Least Sq Mean	Std Error	Lower 95%	Upper 95%
c	70.185714	1.7655175	66.324558	74.046870
i	75.900000	1.7655175	72.038844	79.761156
n	71.100000	1.7655175	67.238844	74.961156

wise treatments differences using either an unadjusted Student's t test for each comparison or using the multiple-comparisons-adjusted Tukey's Honest Significant Difference (HSD) tests for each pair as in Figure 2.4.

Here we see that the pressure required to break the iron bonds is 5.7 units greater, on average, than the pressure required to break the copper bonds. This adjusted HSD test has a standard error of the difference of 1.72. The two-sided 95% confidence interval on this difference is from 1.1 to 10.3. Because this interval does not contain zero, or,

Figure 2.4: Tukey HSDs for Metal Bond Breaking

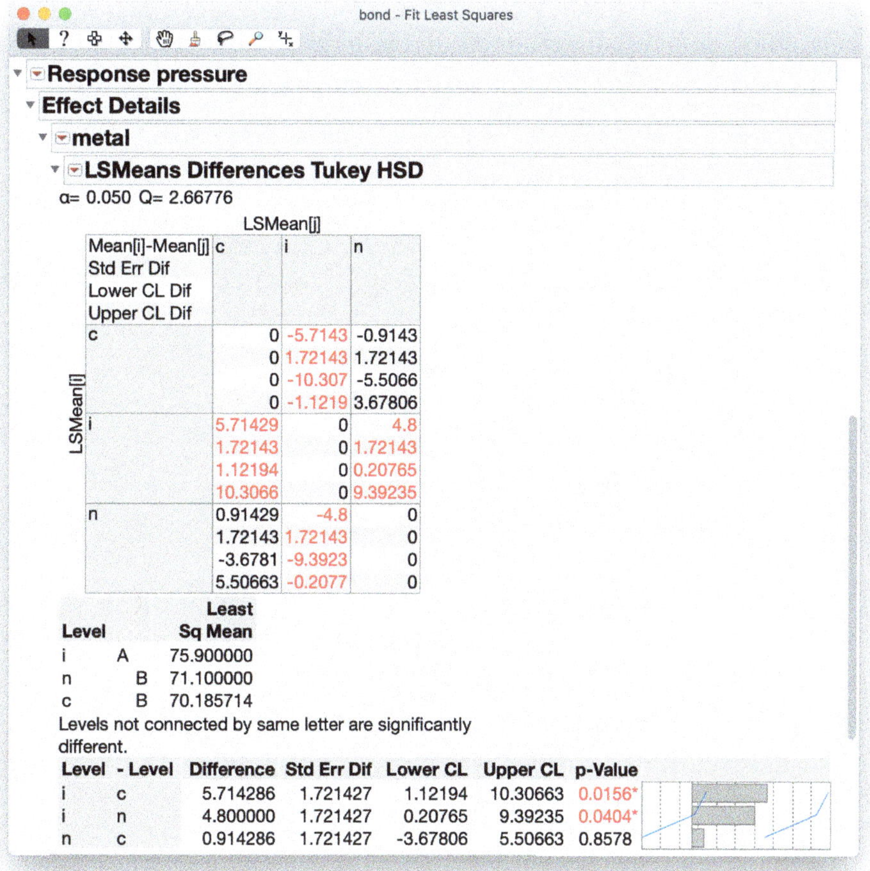

bond - Fit Least Squares

Response pressure

Effect Details

metal

LSMeans Differences Tukey HSD

α= 0.050 Q= 2.66776

LSMean[j]

Mean[i]-Mean[j] Std Err Dif Lower CL Dif Upper CL Dif	c	i	n
c	0 0 0 0	-5.7143 1.72143 -10.307 -1.1219	-0.9143 1.72143 -5.5066 3.67806
i	5.71429 1.72143 1.12194 10.3066	0 0 0 0	4.8 1.72143 0.20765 9.39235
n	0.91429 1.72143 -3.6781 5.50663	-4.8 1.72143 -9.3923 -0.2077	0 0 0 0

LSMean[i]

Level		Least Sq Mean
i	A	75.900000
n	B	71.100000
c	B	70.185714

Levels not connected by same letter are significantly different.

Level	- Level	Difference	Std Err Dif	Lower CL	Upper CL	p-Value
i	c	5.714286	1.721427	1.12194	10.30663	0.0156*
i	n	4.800000	1.721427	0.20765	9.39235	0.0404*
n	c	0.914286	1.721427	-3.67806	5.50663	0.8578

equivalently, since the p-value reported for that test in the Ordered Differences report (available from the red triangle for this LSMeans Differences report) is less than the nominal cutoff of 0.05, we can conclude that there is some statistical evidence that the required breaking pressure for iron truly is different from the required breaking pressure for copper. We caution you, as in the first chapter, against using only p-values to make final conclusions. P-values should be guidelines, and these guidelines should always be used in conjunction with the size of the difference and subject-matter considerations to create an argument for a statistically unusual and practically useful result.

We already saw the fixed-effects parameter estimates and LSMeans; now we explore

Figure 2.5: Ingot BLUPs for Metal Bond Breaking

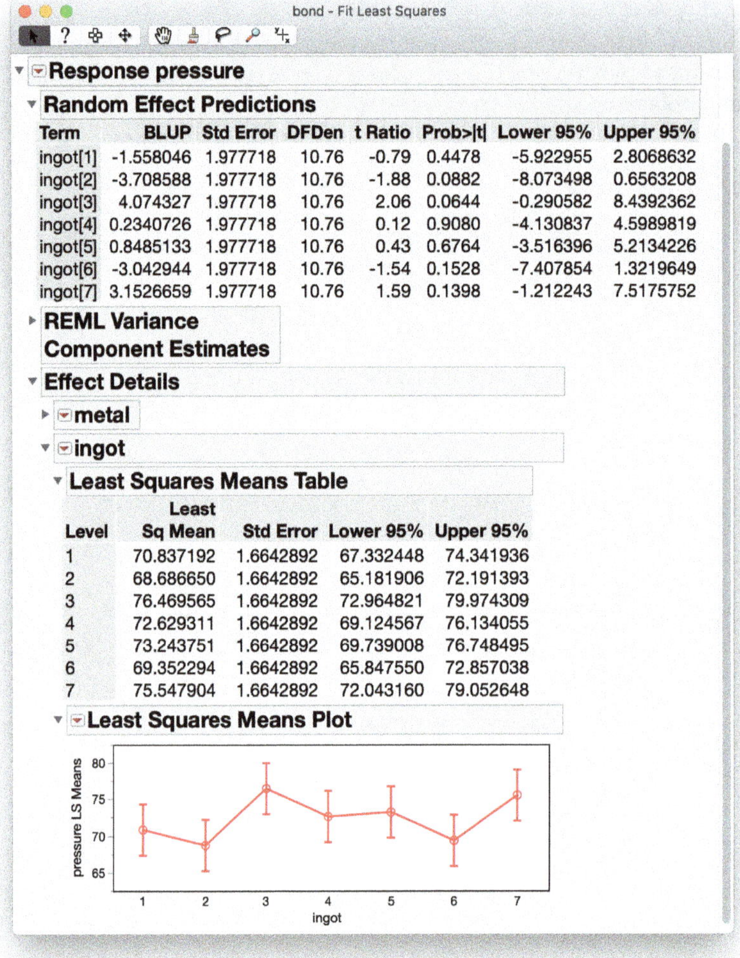

Term	BLUP	Std Error	DFDen	t Ratio	Prob>|t|	Lower 95%	Upper 95%
ingot[1]	-1.558046	1.977718	10.76	-0.79	0.4478	-5.922955	2.8068632
ingot[2]	-3.708588	1.977718	10.76	-1.88	0.0882	-8.073498	0.6563208
ingot[3]	4.074327	1.977718	10.76	2.06	0.0644	-0.290582	8.4392362
ingot[4]	0.2340726	1.977718	10.76	0.12	0.9080	-4.130837	4.5989819
ingot[5]	0.8485133	1.977718	10.76	0.43	0.6764	-3.516396	5.2134226
ingot[6]	-3.042944	1.977718	10.76	-1.54	0.1528	-7.407854	1.3219649
ingot[7]	3.1526659	1.977718	10.76	1.59	0.1398	-1.212243	7.5175752

Level	Least Sq Mean	Std Error	Lower 95%	Upper 95%
1	70.837192	1.6642892	67.332448	74.341936
2	68.686650	1.6642892	65.181906	72.191393
3	76.469565	1.6642892	72.964821	79.974309
4	72.629311	1.6642892	69.124567	76.134055
5	73.243751	1.6642892	69.739008	76.748495
6	69.352294	1.6642892	65.847550	72.857038
7	75.547904	1.6642892	72.043160	79.052648

the corresponding random effect predictions for `ingot`. These are shown as LSMeans in Figure 2.5 and can also be considered as Best Linear Unbiased Predictions (BLUPs) for the ingot random effects.

These are the values that estimate the r_j terms in our statistical model. Note that ingots 3 and 7 have the highest breaking strength effects, and ingots 2 and 6 have the lowest. It might be worth investigating whether there was some defect in any of the ingots if we see extreme effects coming from them. Remember that we are expecting these BLUPs to be normally distributed, centered around zero with constant variance, and in fact the BLUPs themselves are shrunken towards zero as compared to what they would be had we declared ingot to be a fixed effect. The t test statistics and corresponding p-values can provide pointers to any experimental results that seem suspiciously far from zero and require further investigation and possible intervention.

There are even more reporting and modeling options available from the red triangle menu at the top of the model report. We will explore some of these additional options in more detail in other examples throughout this book. You can also consult the JMP documentation to learn more about the statistical details, examples, and options by performing a web search of the option along with "JMP" or going directly to the url for the documentation (SAS Institute Inc. 2019a).

Repeating the Analysis using JMP Pro

Now let's explore the same example using the Mixed Model personality in JMP Pro.

JMP Pro Instructions for Metal Bond Breaking

Use *Analyze > Fit Model* to define the model with one fixed factor and one random factor.
Enter **pressure** as the *Y* variable. Change the default personality from *Standard Least Squares* to *Mixed Model*. Enter **metal** in the *Fixed Effects* tab of the *Construct Model Effects* box.
Enter **ingot** in the *Random Effects* tab of the *Construct Model Effects* box.
Click *Run* to fit the model.

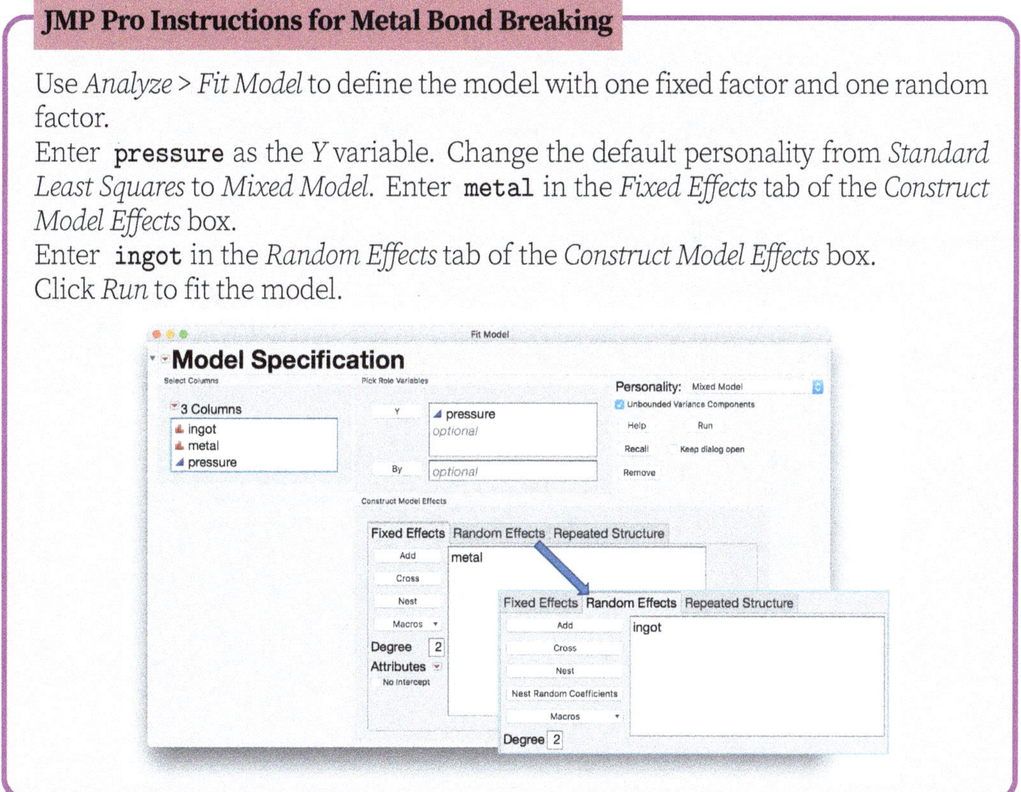

Variance Components: Bounded or Unbounded?

When filling out the dialog for the Mixed Model in JMP Pro, you will notice a selection box next to the option for *Unbounded Variance Components*, which is selected by default. In general, for experiments wherein you have interest in the fixed effects, you should select the *Unbounded Variance Components* option. This allows the estimation of all parameters in the model to be unbiased. It might, at first thought, seem like a benefit to force all variance components to be nonnegative. After all, it doesn't make any sense for a variance to be negative – a variance is a sum of squares of differences – and that can never be negative! But, in fact, the estimation of all of these pieces of the model are connected to each other, and allowing a variance component estimate to stray into the negatives might actually improve the estimation for other components and effects. The situations when this is likely to happen usually involve a weak variance signal for that component, either because the variance truly is quite small or because there are not many levels to that random effect variable, or due to a negative correlation within blocks, as described in Chapter 1.

In summary, choose the *Unbounded Variance Components* option to allow negative variances unless you are only interested in getting variance estimates from all of your random effects (and not in estimating any fixed effects) or you have a good reason to disallow negative estimates. See more detail about this by searching "mixed models and random effects models" in the JMP documentation (SAS Institute Inc. 2019a).

Interpreting Output for JMP Pro's Mixed Model Personality

The output from the Mixed personality, available only in JMP Pro, is quite similar to that from the Standard Least Squares personality, but you will find options in slightly different places as shown in Figure 2.6. For example, the corrected AIC (AICc) and BIC fit statistics are printed by default in this output, but they had to be requested from the top red triangle in the Standard Least Squares output. The LSMeans and comparisons now (for the Mixed personality) have to be requested from the *Multiple Comparisons* option under the red triangle while they were defaults in the Standard Least Squares output.

Figure 2.6: Fit Mixed Output for Metal Bond Breaking

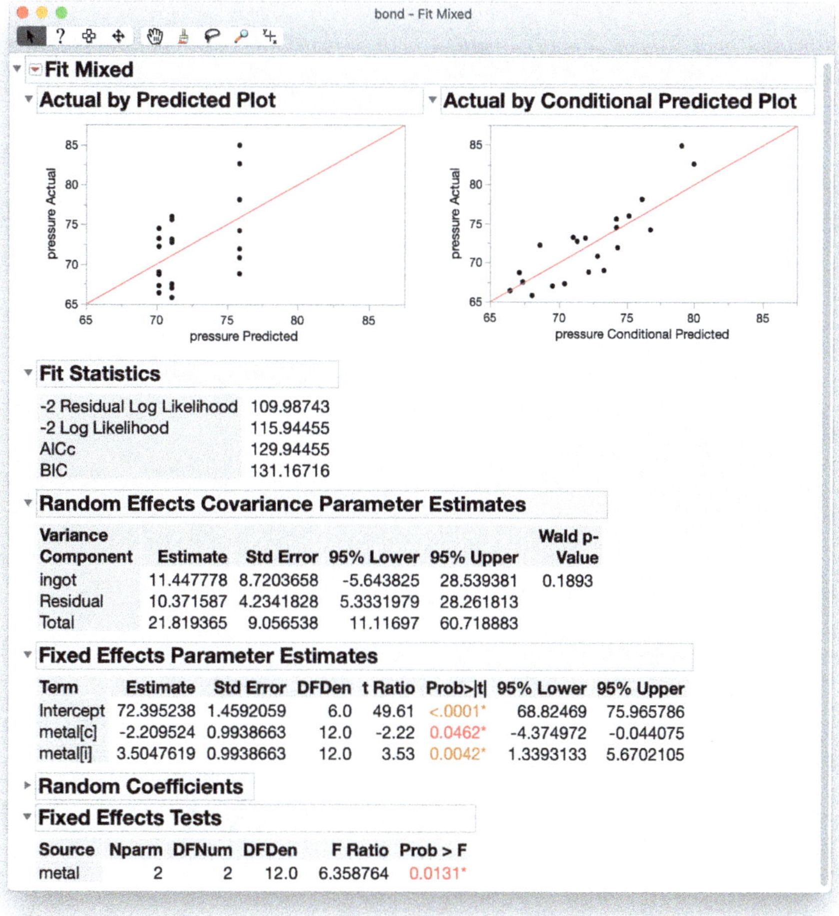

To request the LSMeans and multiple comparisons for the metal applications, choose the *Multiple Comparisons* option under the red triangle to request the LSMeans and comparisons. Choose `metal` (the only option) in the Effects list of the Multiple Comparison window. Check the *Show Least Squares Means Plot* box. Figure 2.7 shows the completed Multiple Comparisons window for the interaction plot and a new display for the pairwise LSMeans comparisons. Hovering your mouse over the dot in the middle of the colored lines prompts a label to appear to identify the metals involved in that comparison (and their estimated difference). If the diagonal line for a comparison does not cross the long diagonal line (indicating no difference), then that comparison is statis-

Figure 2.7: Fit Mixed Output of Tukey HSDs for Metal Bond Breaking

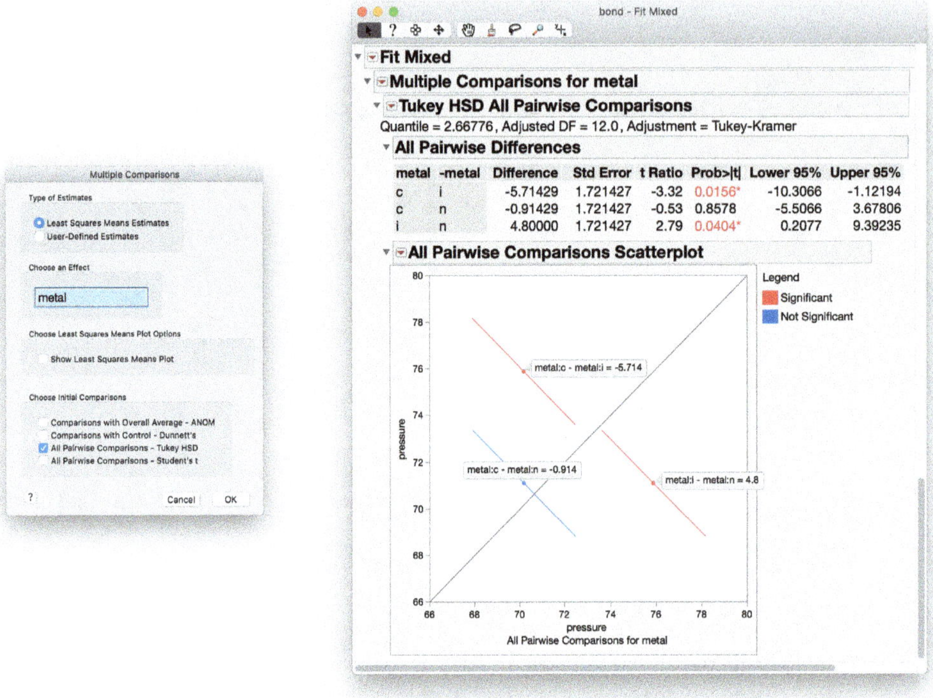

Quantile = 2.66776, Adjusted DF = 12.0, Adjustment = Tukey-Kramer

metal	-metal	Difference	Std Error	t Ratio	Prob>|t|	Lower 95%	Upper 95%
c	i	-5.71429	1.721427	-3.32	0.0156*	-10.3066	-1.12194
c	n	-0.91429	1.721427	-0.53	0.8578	-5.5066	3.67806
i	n	4.80000	1.721427	2.79	0.0404*	0.2077	9.39235

tically significantly different from zero. Those comparison lines also turn colors to indicate statistical significance at the 0.05 level: red indicates a statistically significant difference and blue indicates no statistically significant difference.

2.4 Balanced Incomplete Blocks Example

Sometimes we do not have a sufficient number of experimental units per block to randomize every treatment into every block. For example, a researcher has four blocks of material, and each block can be used for three runs. But say that the researcher has four treatment levels to test. In this case, we can choose to randomize the treatments to the blocks so that each treatment shows up in three of the four blocks. This would be called an *Incomplete Block Design*. If we further require that all of the pairs of treatments show up an equal number of times together, we have a special design called a *Balanced Incomplete Block Design*, or *BIBD*. By pairing the treatments within blocks, we can orthogonalize the block effect from the comparison between the treatments. The data in this example, shown in Figure 2.8, are from Example 4-5 in Montgomery (2005). These data are in a file called `bibd.jmp`.

Figure 2.8: Balanced Incomplete Blocks Data

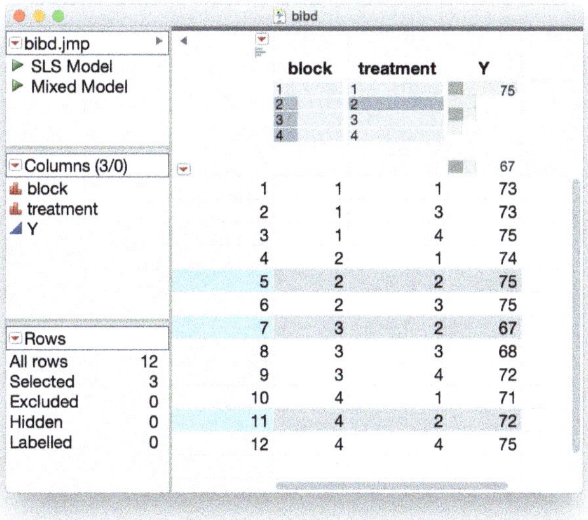

	block	treatment	Y
1	1	1	73
2	1	3	73
3	1	4	75
4	2	1	74
5	2	2	75
6	2	3	75
7	3	2	67
8	3	3	68
9	3	4	72
10	4	1	71
11	4	2	72
12	4	4	75

Here is the Skeleton ANOVA for this example:

Experiment Design		**Skeleton ANOVA**	
Source	***df***	**Source**	***df***
Block	4-1=3	Block	3
		Treatment	4-1=3
Unit	12	Residual	12-3=9
Total	16-1=15	Total	15

We set up the analysis in the same way as we do for an RCBD. The results will also look the same. The only important difference in a BIBD versus and RCBD is that the treatments must be allocated to the blocks properly in advance of collecting the response data. If we allocate the treatments properly, then the analysis will be interpretable in essentially the same way as an RCBD, because the design allows for the estimation of those treatment differences separably from the block estimates.

JMP Instructions for Balanced Incomplete Blocks

Go to *Analyze > Fit Model*. Enter `Y` as the *Y* variable. Note that the default personality becomes *Standard Least Squares* because you've entered a continuous variable in the *Y* role. Keep this default. Enter **`block`** and **`treatment`** in the *Construct Model Effects* box. By default, these are *Fixed Effects*. To make the **`block`** random, select the *Attributes* drop-down and select **`Random Effect`**. Now **`block`** has the "& Random" appearing to the right in the *Construct Model Effects* box. Click *Run* to fit the model.

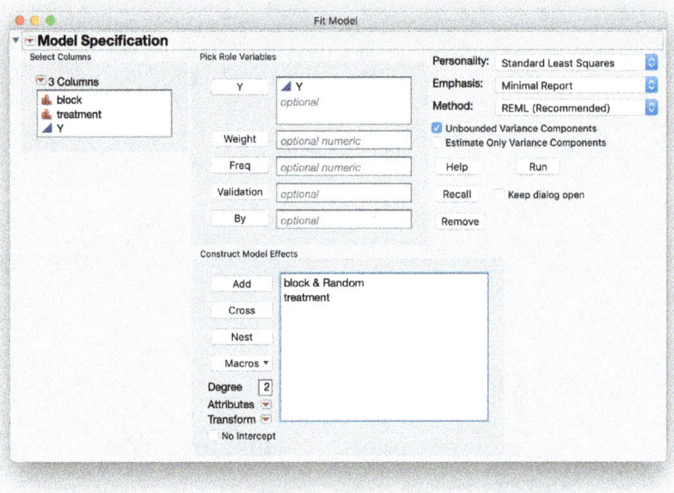

Figure 2.9 shows the primary results, with the Parameter Estimates and Random Effect Predictions expanded, and the Tukey HSD pairwise comparisons. The F test for treatment, with F Ratio 11.33 and *p*-value 0.0112, indicates that there is a statistically significant difference among the treatment groups, but it does not identify which comparison is statistically significantly different.

From the Tukey HSD pairwise comparisons, we see that Treatment 4 is statistically significantly larger than all other treatments, and none of the other treatments have any statistically significant differences among themselves. This trend is also obvious in the Least Squares Means Plot.

Figure 2.9: Results for Balanced Incomplete Blocks

bibd - Fit Least Squares

Response Y

Summary of Fit

RSquare	0.959166
RSquare Adj	0.943853
Root Mean Square Error	0.806226
Mean of Response	72.5
Observations (or Sum Wgts)	12

Parameter Estimates

Term	Estimate	Std Error	DFDen	t Ratio	Prob>|t|
Intercept	72.5	1.434689	2.98	50.53	<.0001*
treatment[1]	-1.086885	0.428359	5.033	-2.54	0.0518
treatment[2]	-0.883607	0.428359	5.033	-2.06	0.0937
treatment[3]	-0.5	0.428359	5.033	-1.17	0.2954

Random Effect Predictions

Term	BLUP	Std Error	DFDen	t Ratio	Prob>|t|
block[1]	0.8491803	1.477019	3.325	0.57	0.6020
block[2]	2.9114754	1.477019	3.325	1.97	0.1343
block[3]	-3.760656	1.477019	3.325	-2.55	0.0761
block[4]	0	1.477019	3.325	0.00	1.0000

REML Variance Component Estimates

Random Effect	Var Ratio	Var Component	Std Error	95% Lower	95% Upper	Wald p-Value	Pct of Total
block	12.333333	8.0166667	6.7463636	-5.205963	21.239296	0.2347	92.500
Residual		0.65	0.4110961	0.2532632	3.909955		7.500
Total		8.6666667	6.7494942	2.8877579	99.87517		100.000

-2 LogLikelihood = 36.993052677

Note: Total is the sum of the positive variance components.

Total including negative estimates = 8.6666667

Covariance Matrix of Variance Component Estimates

Iterations

Fixed Effect Tests

Source	Nparm	DF	DFDen	F Ratio	Prob > F
treatment	3	3	5.033	11.3294	0.0112*

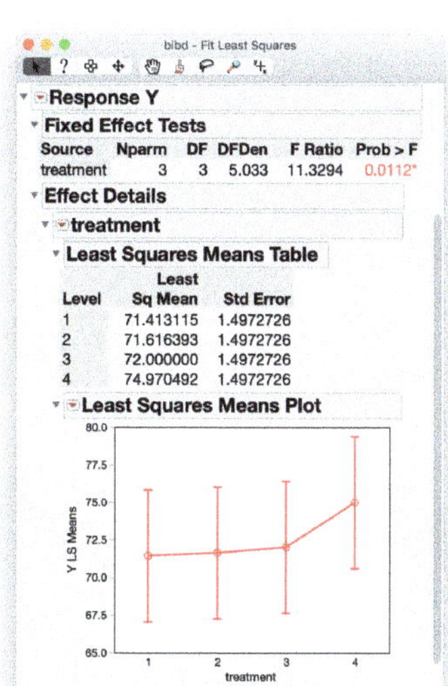

bibd - Fit Least Squares

Response Y

Fixed Effect Tests

Source	Nparm	DF	DFDen	F Ratio	Prob > F
treatment	3	3	5.033	11.3294	0.0112*

Effect Details

treatment

Least Squares Means Table

Level	Least Sq Mean	Std Error
1	71.413115	1.4972726
2	71.616393	1.4972726
3	72.000000	1.4972726
4	74.970492	1.4972726

Least Squares Means Plot

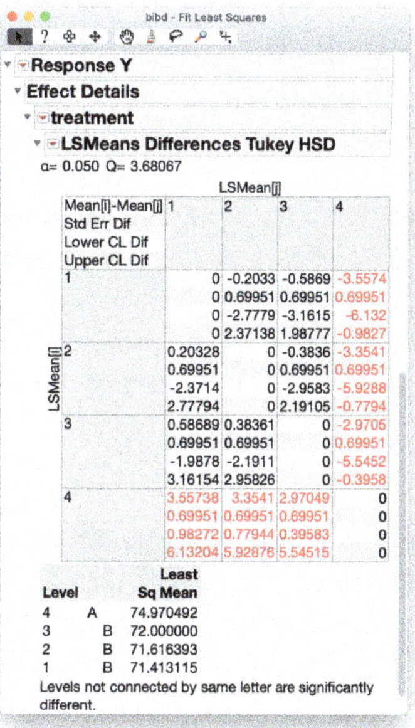

bibd - Fit Least Squares

Response Y

Effect Details

treatment

LSMeans Differences Tukey HSD

α= 0.050 Q= 3.68067

Mean[i]-Mean[j] Std Err Dif Lower CL Dif Upper CL Dif	1	2	3	4
1	0 0 0 0	-0.2033 0.69951 -2.7779 2.37138	-0.5869 0.69951 -3.1615 1.98777	-3.5574 0.69951 -6.132 -0.9827
2	0.20328 0.69951 -2.3714 2.77794	0 0 0 0	-0.3836 0.69951 -2.9583 2.19105	-3.3541 0.69951 -5.9288 -0.7794
3	0.58689 0.69951 -1.9878 3.16154	0.38361 0.69951 -2.1911 2.95826	0 0 0 0	-2.9705 0.69951 -5.5452 -0.3958
4	3.55738 0.69951 0.98272 6.13204	3.3541 0.69951 0.77944 5.92876	2.97049 0.69951 0.39583 5.54515	0 0 0 0

Level			Least Sq Mean
4	A		74.970492
3		B	72.000000
2		B	71.616393
1		B	71.413115

Levels not connected by same letter are significantly different.

Repeating the Analysis using JMP Pro

To run this example in JMP Pro, we do the following.

JMP Pro Instructions for BIBD.jmp Example

Go to *Analyze > Fit Model*. Enter `Y` as the *Y* variable. Change the default personality from *Standard Least Squares* to *Mixed Model*. Enter `treatment` in the *Fixed Effects* tab of the *Construct Model Effects* box. Enter `block` in the *Random Effects* tab of the *Construct Model Effects* box.
Click *Run* to fit the model.

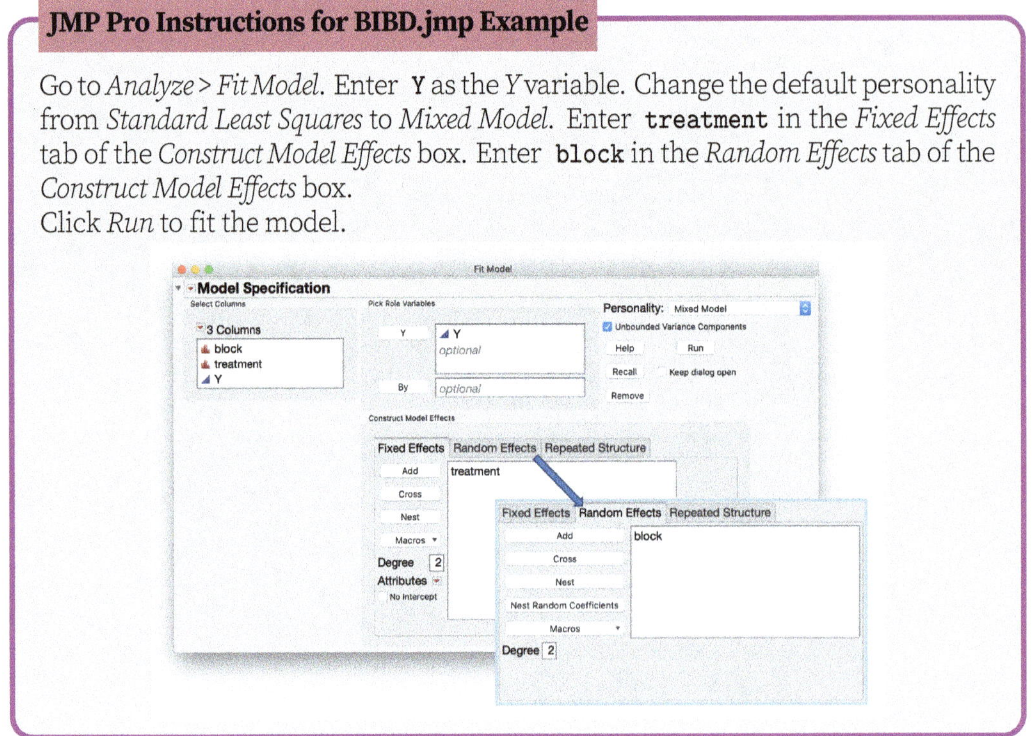

The JMP Pro results in Figure 2.10 are quite similar, with slightly different arrangement of options.

Let's explore the special Marginal and Conditional options available in the JMP Pro output. The Marginal Profiler and Marginal Surface Profiler are shown in Figure 2.11. These interactive plots help us to explore the prediction surface averaged over all of the blocks. The profiler shows us the predicted mean response at the various levels of the treatment, along with a 95% confidence interval for that mean response at that treatment level. The surface profiler extends this 2-dimensional plot into a 3-dimensional plot. In this case, with a response and only one fixed effect, we only really need 2 dimensions, so the 3-dimensional plot looks a bit like a staircase staying constant over the block variable and rising over the treatment variable.

The Conditional Profiler and Conditional Surface Profiler are shown in Figure 2.12. Now we can interact with the treatment and the blocks to see how the predictions for the response differ for different treatments when we condition on a specific block. We no longer have confidence intervals for these conditional means. The surface profiler be-

Figure 2.10: JMP Pro Results for Balanced Incomplete Blocks

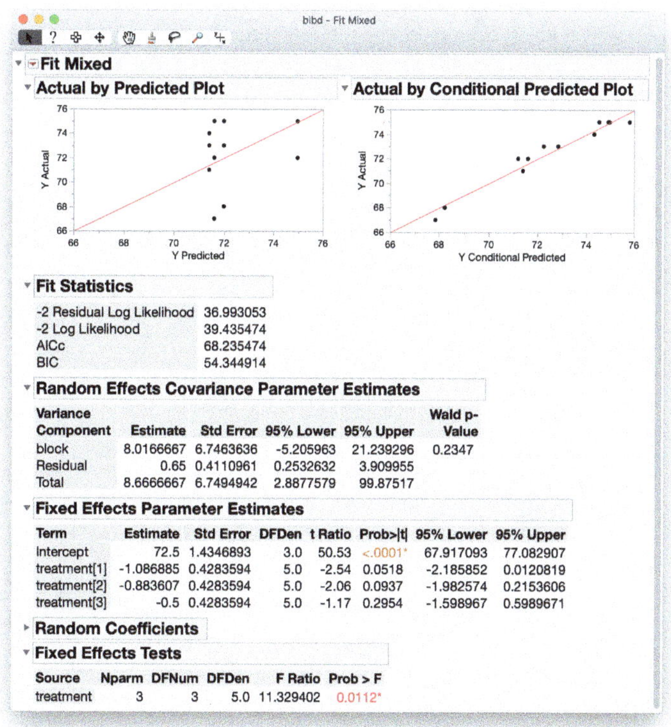

Figure 2.11: Marginal Profiles for Balanced Incomplete Blocks

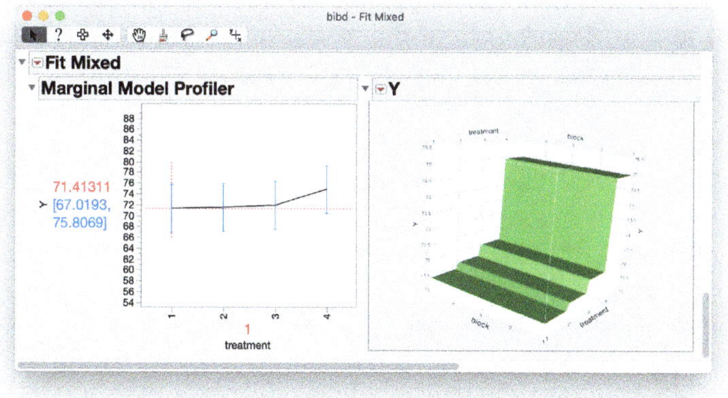

Figure 2.12: Conditional Profiles for Balanced Incomplete Blocks

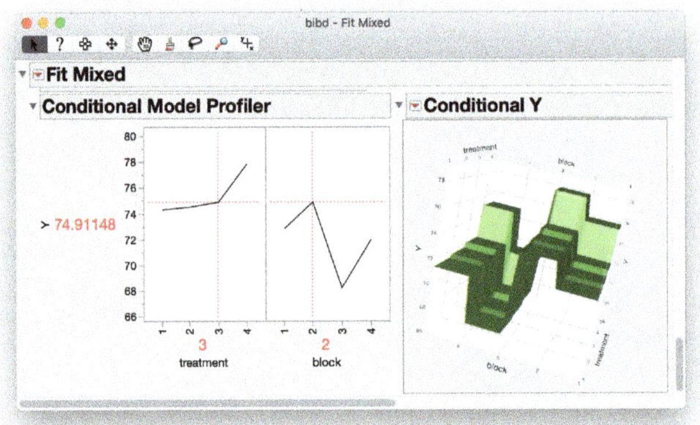

comes more interesting here, because we now have a third dimension, block, to plot along with the response and the treatment.

Summary

In this chapter, we have explored including a single blocking variable as a random effect in a model, with various amounts of imbalance or missingness accounted for automatically by using the REML method of estimation. We will explore a few further examples of experiments with a single blocking variable in the exercises. In the next chapter, we will extend the treatment structure to include factorial designs, and we will begin to examine more complex blocking structures as well.

2.5 Exercises

1. Cell Viability Example (as described in Section 1.2).
   (a) Analyze this experiment assuming that the blocks (Plates) are fixed. Report the LSMeans for the `Chemical` treatments, and report the difference, standard error of the difference, the 95% Confidence Interval of the difference, and the p-value of the test of the difference between the Chemical treatments.
   (b) Analyze this experiment assuming that the blocks (Plates) are random. Report the LSMeans for the `Chemical` treatments, and report the difference, standard error of the difference, the 95% Confidence Interval of the difference, and the p-value of the test of the difference between the Chemical treatments.
   (c) How do the results differ?
   (d) Which analysis is more appropriate for this scenario? Why?
2. RCBD with interaction replication: Machine and Operator Example.

   These data are in a file called `MachineBalanced.jmp`. The director of operations at a small manufacturing company suspects that the three machines in her machine shop are producing at different levels of quality. The company develops a continuous part quality rating scale, from 0 to 100, and she wants to rate 18 randomly selected parts produced by each machine (for a total of 54 parts). Six machine operators will assess the ratings of the parts. Each operator will rate three parts from each of the three machines. The order in which the parts are chosen and assigned to the operators for rating is random. The resulting data table is shown below. These data are simulated, but inspired by a similar example in McLean et al. (1991).

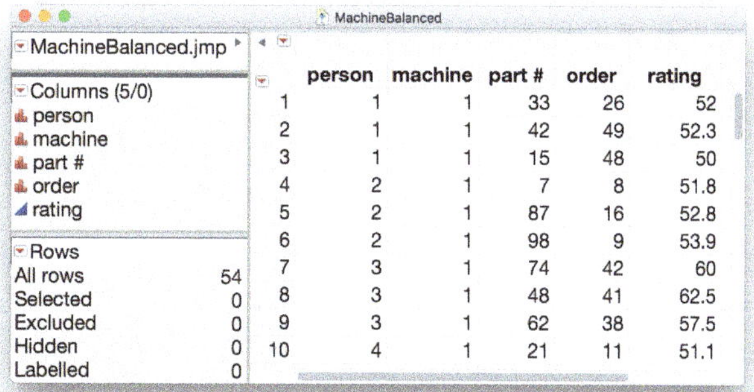

MachineBalanced

MachineBalanced.jmp

Columns (5/0)
person
machine
part #
order
rating

Rows
All rows 54
Selected 0
Excluded 0
Hidden 0
Labelled 0

	person	machine	part #	order	rating
1	1	1	33	26	52
2	1	1	42	49	52.3
3	1	1	15	48	50
4	2	1	7	8	51.8
5	2	1	87	16	52.8
6	2	1	98	9	53.9
7	3	1	74	42	60
8	3	1	48	41	62.5
9	3	1	62	38	57.5
10	4	1	21	11	51.1

In this case, we care about knowing how the Machines differ in performance (**Ratings**), but the **Person** is more of a nuisance variable. Our primary use of **Person** in this design was to hurry up the execution of the experiment (more raters means we can finish the experiment sooner) and to make sure that we didn't have bias from using only one rater whose consistency is unknown. We only want to use the **Person** differences to explain the variation in the experiment that comes from **Person**.

This is another example of an *RCBD*. The idea is that we blocked by Operator, and we randomly assigned the three Machines equally to all six Operators. However, instead of only assigning each **Machine** to each **Person** for a single trial, we replicated the **Person*Machine** for a total of three experimental units. The **bond** example was an unreplicated RCBD, and this **machine** example is a replicated RCBD.

(a) Write out the statistical model for this example.
(b) Fit this model in JMP or JMP Pro. How much variability in the Ratings is explained by the Operator effect? Interpret the results to determine if the three machines in her machine shop are producing at different levels of quality.
(c) What other sources of "batch" effects might be present in these data? For example, think about the time of collection.
(d) If the parts were randomly sampled on different days (over three days total for the entire experiment), how could this effect be addressed in the statistical model?
(e) If each of the three measurements on a particular part by a particular operator were taken on different days, what model would you fit? (Hint: make a new column in the data table for "day" with values 1, 2, and 3. Try building your model

in JMP.) Which factors are random and which are fixed? Can you include all of the interaction terms? How does the sample size limit your model? How does JMP warn you about a problem with fitting the full model when there is no replication?

3. Unbalanced Machine and Operator Example.

   These data are in a file called **`Machine.jmp`**. In this example, we have the same scenario as in the previous example, but we were unable to collect information with equal replication from each operator. Each operator did evaluate at least one part from each machine, but the number of replicates varies by operator. This is called an *unbalanced* design, and we will explore how having imbalance or missing data can impact a mixed model analysis.

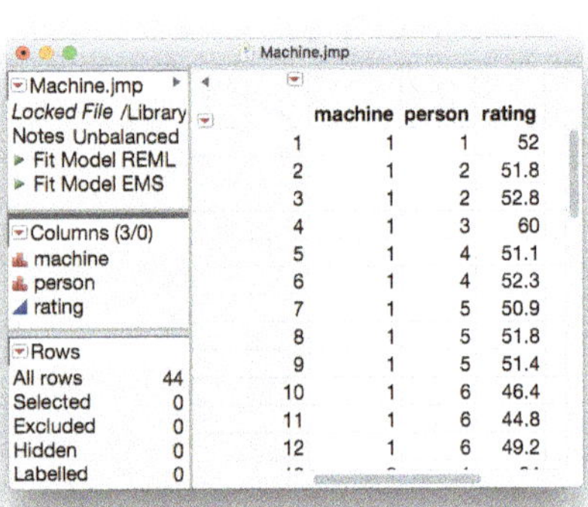

	machine	person	rating
1	1	1	52
2	1	2	51.8
3	1	2	52.8
4	1	3	60
5	1	4	51.1
6	1	4	52.3
7	1	5	50.9
8	1	5	51.8
9	1	5	51.4
10	1	6	46.4
11	1	6	44.8
12	1	6	49.2

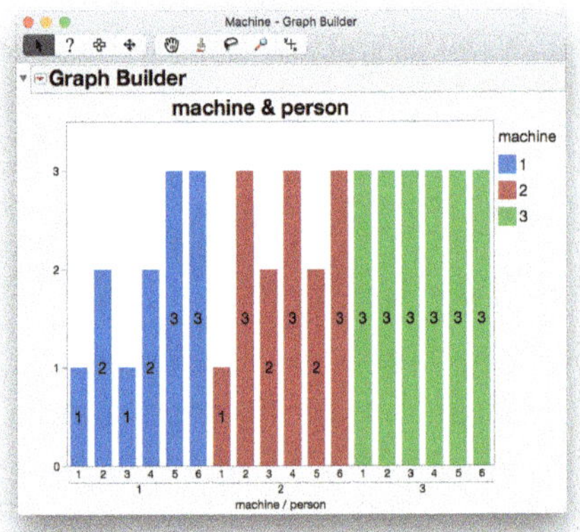

   (a) Write out the statistical model for this example.
   (b) Fit this model in JMP or JMP Pro. How much variability in the Ratings is explained by the Operator effect? Interpret the results to determine if the three machines in her machine shop are producing at different levels of quality.
   (c) Compare these results to the results in the previous exercise (which had no missing data).

4. PBIBD: Cotton Example.

   These data are in a file called **`cotton.jmp`**. In this example, from Cochran and Cox (1957), an experimenter is comparing the yield, in pounds of seed cotton per plot, of fifteen field treatments. The experimenter only has fifteen strips of field to use for the experiment, and each strip can be divided into four plots on which to apply different treatments. So, in total, the experimenter can use 15 strips with four experimental units, which is 60 experimental units total. The fifteen treatments can each be used four times. Because the experimenter cannot use each treatment in each block, we do not have a full RCBD. We might be looking at a BIBD, but upon further investigation we realize that this design does not allow all pairs of treatments to show up together in a block. (If you open the data table and explore the design, you can see that, while treatments 1 and 2 appear together in the third block, treatments 1 and 6 never appear together in any block.) The data table is shown here along with bar charts of the **block** and **treatment** factors and a histogram of the **pounds** response.

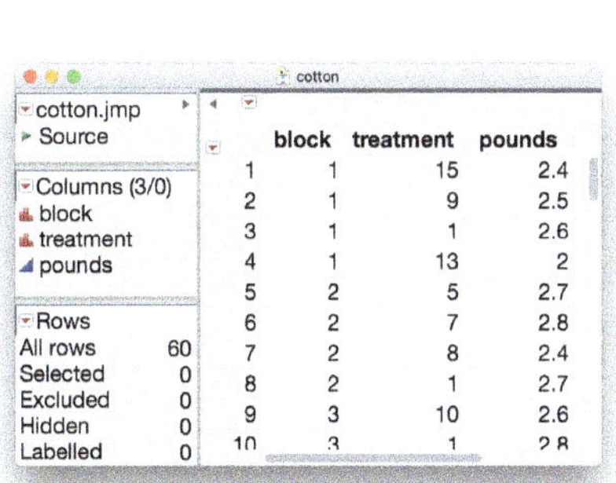

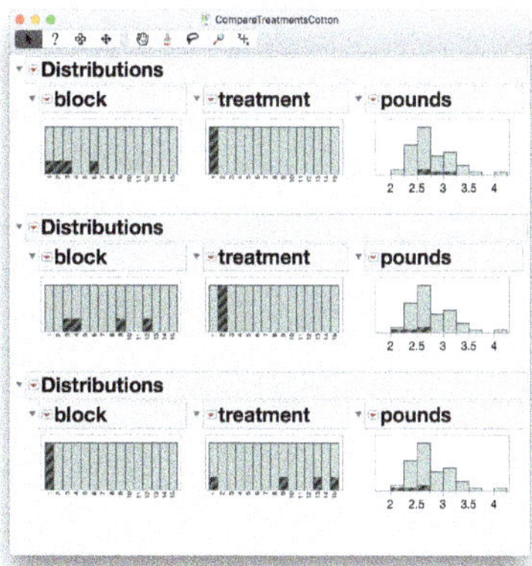

This is a *partially balanced incomplete block design*, or *PBIBD*.

Again, we have a treatment variable, **treatment**, which is of primary interest for comparisons and should be treated as fixed. We also have a blocking variable, **block**, which is not of interest for comparisons and can be treated as a random sample of possible blocks.

(a) Write out the statistical model for this example.

(b) Fit this model in JMP or JMP Pro. How much variability in the pounds of seed cotton per plot is explained by the block effect? Interpret the results to determine if the fifteen treatments differ in their mean response, and, if so, which treatment(s) is/are the best (or are not statistically significantly different from the best).

Chapter 3

Models with Factorial Treatment Designs

Our focus in this chapter bends slightly away from mixed models directly to investigate what is arguably the most common treatment design used in conjunction with mixed models —the factorial treatment design. We will start with an introduction to factorial designs in a simple randomized complete block experiment. Then we will bring ourselves back to mixed model territory with the split-plot design. The factorial treatment design in conjunction with variations of the basic split-plot experiment design is probably the most common form of mixed model.

3.1 Motivating Examples

Tensile Strength — A fabric manufacturer wants to test the tensile strength of a fabric after washing under various conditions. She has three machines available each with a setting for temperature, hot and cold, and for water level, high and low. It is possible that both temperature and water level affect the strength of the fabric.

Greenhouse — A plant researcher has two plant varieties and a pesticide meant to protect the plants against disease. In the greenhouse, the pesticide can only be applied to large sections of the benches. The bench sections can hold multiple plants. The researcher wants to identify the best plant and pesticide combination for disease resistance.

Semiconductor — A semiconductor engineer has several process conditions that affect resistance on the wafers produced. The position of the chips on the wafer can also affect resistance. The engineer wants to minimize resistance and understand the effects of process condition and position.

3.2 Conceptual Background

The examples discussed in Chapter 2 all had a single treatment factor of interest. Many times researchers have more than one treatment factor that they can control that could affect the response. Those researchers *could* run multiple experiments, one for each factor, but this is inefficient in both resources (time, equipment, etc.) and statistical efficiency. Specifically, when the experiments are run separately, the possibility of *in-*

teraction between the treatment factors cannot be tested.

Key Terminology

Main Effect The effect of a single factor averaging over any other factors in the model.

Simple Effect The effect of one factor given a particular level of another factor.

Interaction An interaction occurs when the effect of one factor changes depending on the level of another factor. If the *simple effect* of a factor is different depending on which level of the other factor is used, then there is an interaction between the two factors.

The potential for interaction between two or more factors means we should design experiments to measure the interaction. The presence of an interaction typically restricts any further treatment comparisons tested after the initial interaction test. We will see how this works in the examples.

When there are two or more treatment factors, and one is harder to change than the other, a particular experiment design is natural. This design is the split-plot experiment design (and all of its more complicated relatives, the split-split plot, the split-strip plot, etc.).

In the split-plot design, the hard-to-change factor is assigned to the larger experimental unit, known as the whole plot. This assignment can be done either in a completely randomized manner (in a CRD) or when there is a need to block, in some form of a blocked design (RCBD, BIBD, etc.). Each of the whole plot units are then subdivided into split-plot units to which the easy-to-change factor is assigned.

The greenhouse example is a true split-plot. The sections of the greenhouse benches are whole plots to which the amount of pesticide is applied. Within those sections, the placement of the two plant varieties is randomized to the "split" section.

In the semiconductor example described above, process conditions can only be applied to whole wafers. Therefore wafer is the whole plot experimental unit. The individual chips are physical subdivisions of the whole wafer and thus split the whole plot. The position treatment is assigned to the split-plot chips. Technically speaking, position is not a true split-plot effect as position is the same on all wafers and not randomized. The analysis, however, is the same.

3.3 RCBD with Factorial Treatments, Tensile Strength Example

The following example explores the tensile strength of fabric under differing care conditions. These data are in a file called `Tensile Strength.jmp`. The manufacturer has three washing machines available for this test. Each machine has settings for `Temperature`, cold and hot, and `Water` level, low and high. Both temperature and water level could affect the measured strength of the fabric, and there may be variability as a result of the particular machine used.

Each washing machine can have only one setting for each of temperature and water level for a particular run. Each machine can be run four times in the time allotted for the experiment. This naturally creates blocks of size four to which the combinations of temperature and water level can be randomized to the runs. This is an RCBD with factorial treatments. Later in this chapter we will see a similar experiment design that requires a different analysis to correctly capture the way the treatments are randomized to the experimental units, so it is worth noting now how we can identify that we have an RCBD analysis and not a split-plot analysis in this example. In this case, we can see that it is the combination of the temperature and water level that are completely randomized to ANY of the four runs within a machine. There is no non-randomness to these treatment applications, except that all combinations must be present in all blocks. The blocks are the only feature restricting our randomization of the treatment application. In later examples in this chapter, we will see that the application of one treatment happens to *groups* of experimental units within blocks, and then the other treatment is applied randomly *within* those groups. This is the key difference to notice that will drive an analysis for RCBD with factorial treatments rather than the more complicated analysis of a split-plot. Let's look at the ANOVA breakdown of our RCBD with factorial treatments.

Experiment Design		**Skeleton ANOVA**	
Source	***df***	**Source**	***df***
Machine	3-1=2	Machine	2
		Temperature	2-1=1
		Water	2-1=1
		Temperature*Water	1
Run(Machine)	(4-1)*3=9	Residual	9-3=6
Total	12-1=11	Total	11

Machine is the block and within each block we perform four runs. The notation Run(Machine) is read "run within machine" and is an example of nesting of experimental units. Temperature and Water are the two main fixed effects, each with two levels, and we call this a 2×2 factorial. We write interaction effects using an asterisk (*), and Temperature*Water has a single degree of freedom obtained by multiplying the degrees of freedom of its constituent main effects. The statistical model based on the skeleton ANOVA above is as follows.

Model for the RCBD with Factorial Treatments, Tensile Strength

$$y_{ijk} = \mu + \alpha_i + \beta_j + \alpha\beta_{ij} + r_k + e_{ijk}$$

y_{ijk} is the response of the ith temperature with the jth water level in the kth machine.
μ is the intercept.
α_i is the effect of the ith level of Temperature.
β_j is the effect of the jth level of Water level.
$\alpha\beta_{ij}$ is the effect of the ith Temperature with the jth Water level.
r_k is the effect of the kth random block and $r_k \sim N(0,\sigma_r^2)$.
e_{ijk} is the residual error and $e_{ijk} \sim N(0,\sigma^2)$.

With our statistical model defined we can fit this model in JMP.

JMP Instructions for Tensile Strength

Go to *Analyze > Fit Model.*
Enter `y` as the *Y* variable. Note that the default personality becomes *Standard Least Squares* because you've entered a continuous variable in the *Y* role. Keep this default.
Enter `Machine`, `Temperature` and `Water` in the *Construct Model Effects* box.
Add the interaction of the `Temperature` and `Water` by selecting both from the list of Columns at the same time, and then click the *Cross* button to add this interaction to the *Construct Model Effects* box.
By default, these are all *Fixed Effects*. To make the `Machine` random, click on that variable in the *Construct Model Effects* box, select the *Attributes* drop-down, and select *Random Effect*. Click *Run* to fit the model.

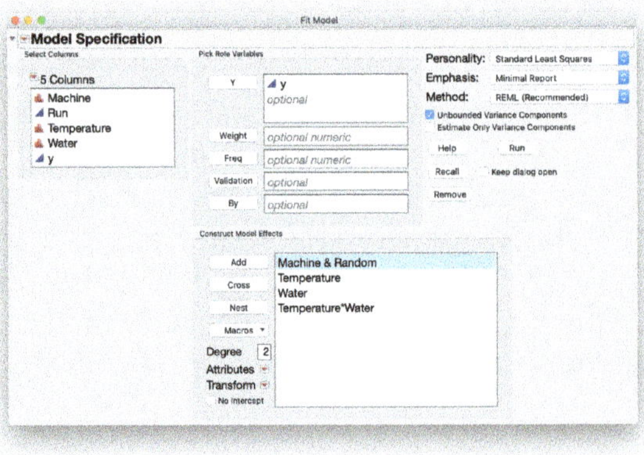

Results and Interpretation

Looking first at the REML Variance Component Estimates in Figure 3.1 we see that the variance for Machine is approximately half the residual variance. The variance is also not significantly different from zero according to the Wald test. This is not concerning as we did not have very many machines available to estimate this variance, and the washing machine manufacturer would probably be glad to see that there is not much variability between machines.

The Fixed Effect Tests show that all of the fixed effects in the model, Temperature, Water and their interaction, are statistically significant. We can also verify that the degrees of freedom agree with our skeleton ANOVA table. Because the interaction effect is significant, the main effect tests for Water and Temperature should be interpreted very cautiously as they are effectively averaging over heterogeneous states. To break down the effects water and temperature have upon tensile strength, we look at simple effects. The next steps show what an interaction looks like graphically and how to investigate simple effects.

Figure 3.1: Initial Results for Tensile Strength

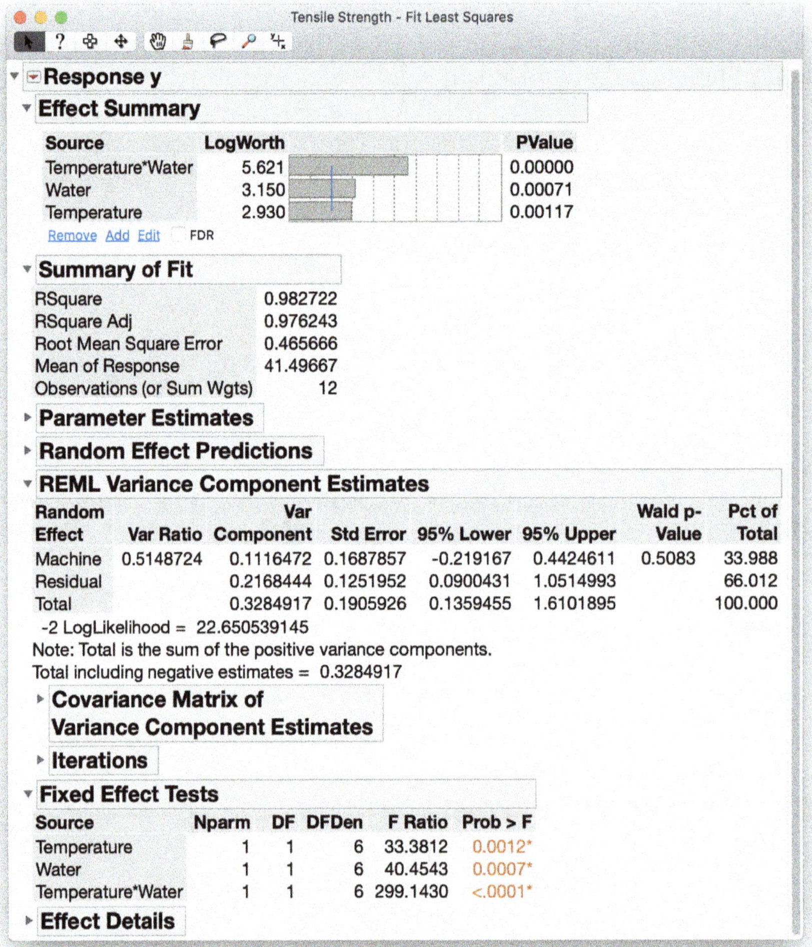

Tensile Strength - Fit Least Squares

Response y

Effect Summary

Source	LogWorth	PValue
Temperature*Water	5.621	0.00000
Water	3.150	0.00071
Temperature	2.930	0.00117

Remove Add Edit FDR

Summary of Fit

RSquare	0.982722
RSquare Adj	0.976243
Root Mean Square Error	0.465666
Mean of Response	41.49667
Observations (or Sum Wgts)	12

Parameter Estimates

Random Effect Predictions

REML Variance Component Estimates

Random Effect	Var Ratio	Var Component	Std Error	95% Lower	95% Upper	Wald p-Value	Pct of Total
Machine	0.5148724	0.1116472	0.1687857	-0.219167	0.4424611	0.5083	33.988
Residual		0.2168444	0.1251952	0.0900431	1.0514993		66.012
Total		0.3284917	0.1905926	0.1359455	1.6101895		100.000

-2 LogLikelihood = 22.650539145

Note: Total is the sum of the positive variance components.

Total including negative estimates = 0.3284917

Covariance Matrix of Variance Component Estimates

Iterations

Fixed Effect Tests

Source	Nparm	DF	DFDen	F Ratio	Prob > F
Temperature	1	1	6	33.3812	0.0012*
Water	1	1	6	40.4543	0.0007*
Temperature*Water	1	1	6	299.1430	<.0001*

Effect Details

Interaction Plot of Temperature*Water Effect

To explore an interaction effect, expand the *Effect Details* report section, scroll down to the interaction effect, and select the LSMeans Plot for Temperature*Water from the red triangle menu. Check the box for *Create an Interaction Plot* and choose with `Temperature` as the overlay.

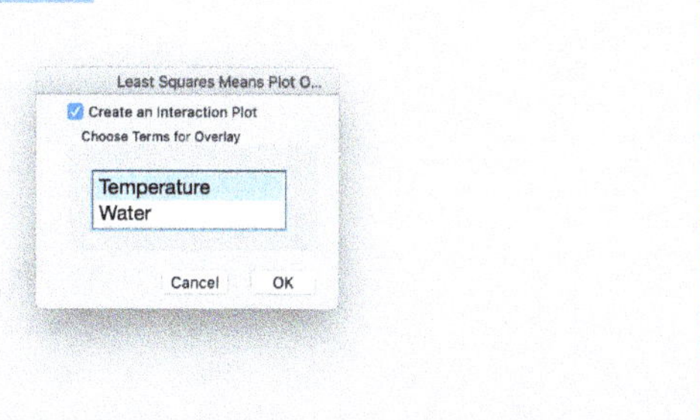

The Least Squares Means Plot in Figure 3.2 shows a classic interaction plot. When the water level is low, hot temperature results in higher tensile strength than cold. However, when the water level is high, the hot temperature results in a lower tensile strength than cold. These are the simple effects of temperature given a particular value of water level. Because the simple effects are different, there is an interaction.

Figure 3.2: Interaction Plot of Temperature*Water for Tensile Strength

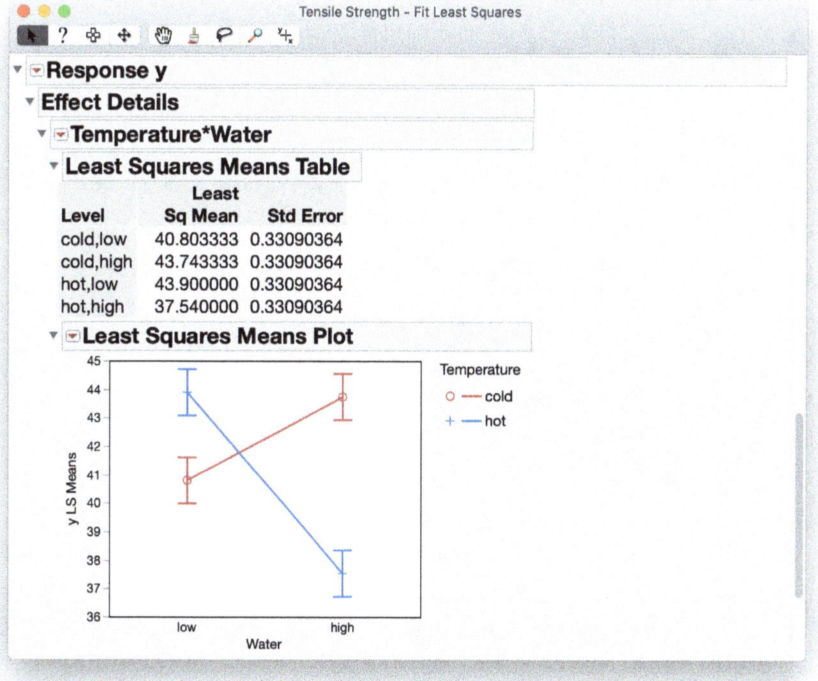

Level	Least Sq Mean	Std Error
cold,low	40.803333	0.33090364
cold,high	43.743333	0.33090364
hot,low	43.900000	0.33090364
hot,high	37.540000	0.33090364

Figure 3.3: Results of Slices of Temperature*Water in Tensile Strength

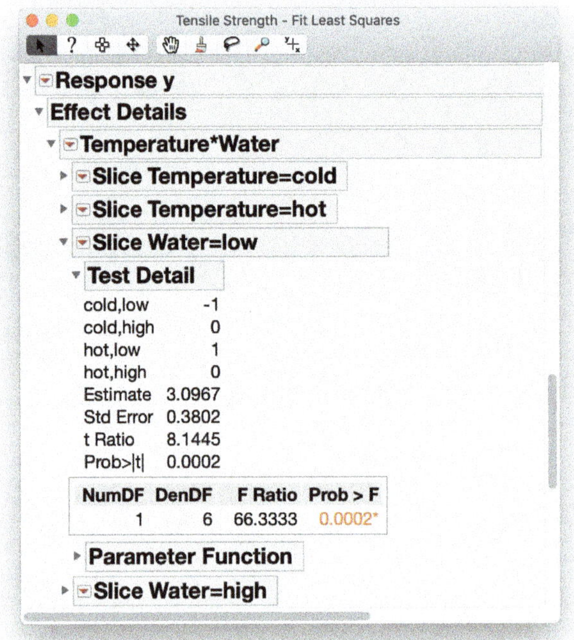

We see this with the crossing lines in the plot and is why the F test for Temperature*Water has a small p-value.

Based on the visual interaction plot and statistical results thus far, we would make the following preliminary conclusions for tensile strength. If water level is low, hot temperature will result in a higher tensile strength. If water level is high, then cold temperature will result in a higher tensile strength. To obtain formal statistical tests and estimates of the simple effects, choose *Test Slices* from the red triangle menu next to Temperature*Water. Results of these slices are shown in Figure 3.3 with the Test Detail where Water=low is expanded.

The formal tests confirm what we see in the interaction plot that all of the conditional slices are statistically significant. Looking specifically at the slice where the water level is held at low, we see that the difference between the hot and cold temperature tensile strengths is 3.0967 units with a standard error for that comparison of 0.38. We know that the hot temperature tensile strength is higher than the cold because `hot,low` has a coefficient of 1 and `cold,low` has a coefficient of -1. Similar information is available for the other slices.

The three unit difference was *statistically* significant. The researcher would be able to say whether this magnitude of difference is *practically* significant. Presumably, this experiment was designed for power to detect at least a certain difference, and this difference is both statistically and practically significant. We will discuss power and how to determine sample size in Chapter 8.

Note that the slices are not adjusted for multiple comparisons. You should typically have in mind a limited number of interesting slices before conducting this test in order to limit the familywise error inflation. When you are interested in many or all of the pairwise comparisons, the Tukey's HSD option will appropriately adjust for multiple comparisons. Remember, though, that these p-values (adjusted or unadjusted) should not be used as hard rules but rather as pieces of evidence for a difference that should always be reported along with the size of that observed difference and an acknowledgement of the size of a difference that would actually be practical and important. However, in the presence of interactions, some of the comparisons in the Tukey's HSD report are not considered valid comparisons. As soon as there is an interaction, any comparisons are restricted to where one of the factors is held at a particular level. In this example, comparing the LSMeans for the cold, low combination to the hot, high combination is disallowed because both factors are changing.

JMP Pro Instructions for Tensile Strength

Go to *Analyze > Fit Model*. Enter **`y`** as the *Y* variable. Change the default personality from *Standard Least Squares* to *Mixed Model*.
Select **`Temperature`** and **`Water`** in the *Select Columns* box then choose *Full Factorial* from the *Macros* list to add both fixed effects and the interaction to the *Construct Model Effects* box.
Enter **`Machine`** in the *Random Effects* tab of the *Construct Model Effects* box. Click *Run* to fit the model.

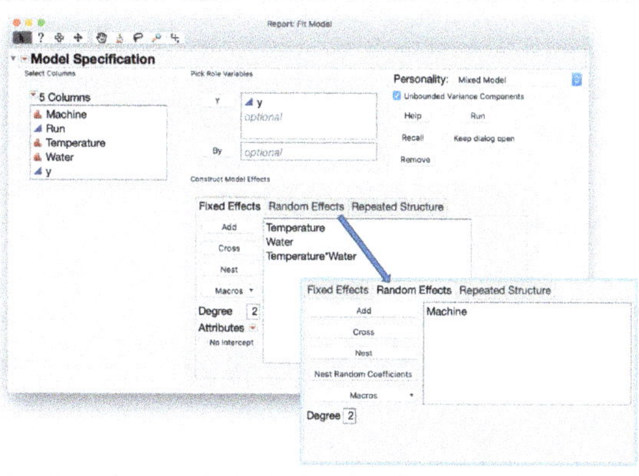

Figure 3.4: Fit Mixed Analysis of Tensile Strength

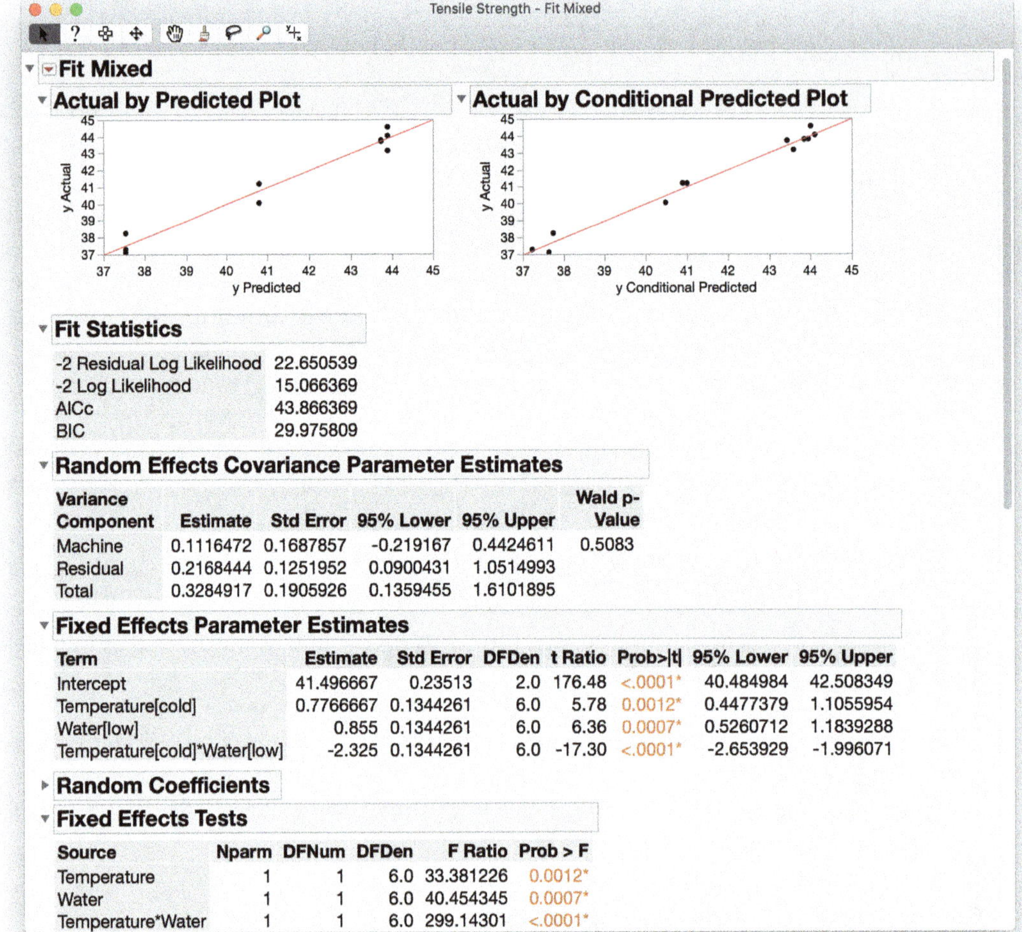

Fit Statistics

-2 Residual Log Likelihood	22.650539
-2 Log Likelihood	15.066369
AICc	43.866369
BIC	29.975809

Random Effects Covariance Parameter Estimates

Variance Component	Estimate	Std Error	95% Lower	95% Upper	Wald p-Value
Machine	0.1116472	0.1687857	-0.219167	0.4424611	0.5083
Residual	0.2168444	0.1251952	0.0900431	1.0514993	
Total	0.3284917	0.1905926	0.1359455	1.6101895	

Fixed Effects Parameter Estimates

Term	Estimate	Std Error	DFDen	t Ratio	Prob>|t|	95% Lower	95% Upper
Intercept	41.496667	0.23513	2.0	176.48	<.0001*	40.484984	42.508349
Temperature[cold]	0.7766667	0.1344261	6.0	5.78	0.0012*	0.4477379	1.1055954
Water[low]	0.855	0.1344261	6.0	6.36	0.0007*	0.5260712	1.1839288
Temperature[cold]*Water[low]	-2.325	0.1344261	6.0	-17.30	<.0001*	-2.653929	-1.996071

Random Coefficients

Fixed Effects Tests

Source	Nparm	DFNum	DFDen	F Ratio	Prob > F
Temperature	1	1	6.0	33.381226	0.0012*
Water	1	1	6.0	40.454345	0.0007*
Temperature*Water	1	1	6.0	299.14301	<.0001*

Results and Interpretation

The Fit Mixed results are the same as the Standard Least Squares personality analysis but include visuals of model fit in the Actual by Predicted and Actual by Conditional Predicted Plots. There are no obvious problems in the data based on these plots shown in Figure 3.4.

A key difference in using the Mixed Model personality comes when investigating the interaction effect further. We use the Multiple Comparisons option rather than having the *Effect Details* report section that the Standard Least Squares personality provides.

You can visualize the interaction plot in Fit Mixed using the Multiple Comparisons tool. *Estimates > Multiple Comparisons* under the red triangle menu brings up the Multiple Comparisons dialog box. Select Temperature*Water, check the Show Least Squares Means Plot box, check the Create an Interaction Plot box, and choose Temperature as the overlay term. The completed dialog box is shown in Figure 3.5.

Figure 3.5: Multiple Comparisons Dialog Box for Interaction Plot

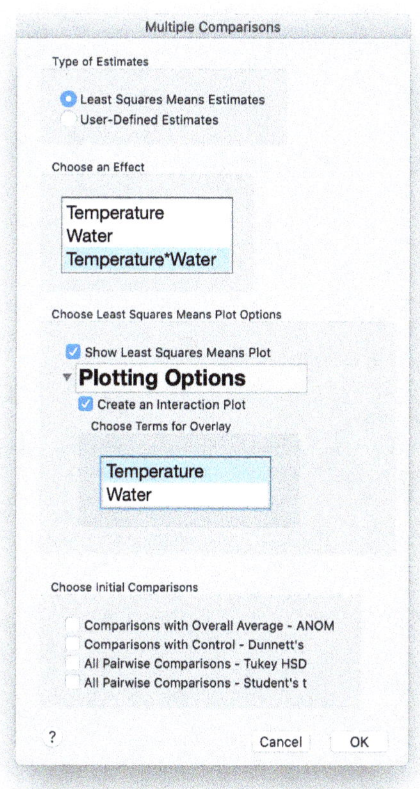

After clicking *OK*, the interaction Least Squares Means Plot is added to the report. The plot is identical to before. The Multiple Comparisons tool in Fit Mixed does not have the same Slice capabilities that Standard Least Squares has, so within this personality we must carefully use the Tukey HSD pairwise comparisons available in the red triangle menu. This is shown in Figure 3.6.

Figure 3.6: Fit Mixed Analysis of Interaction in Tensile Strength

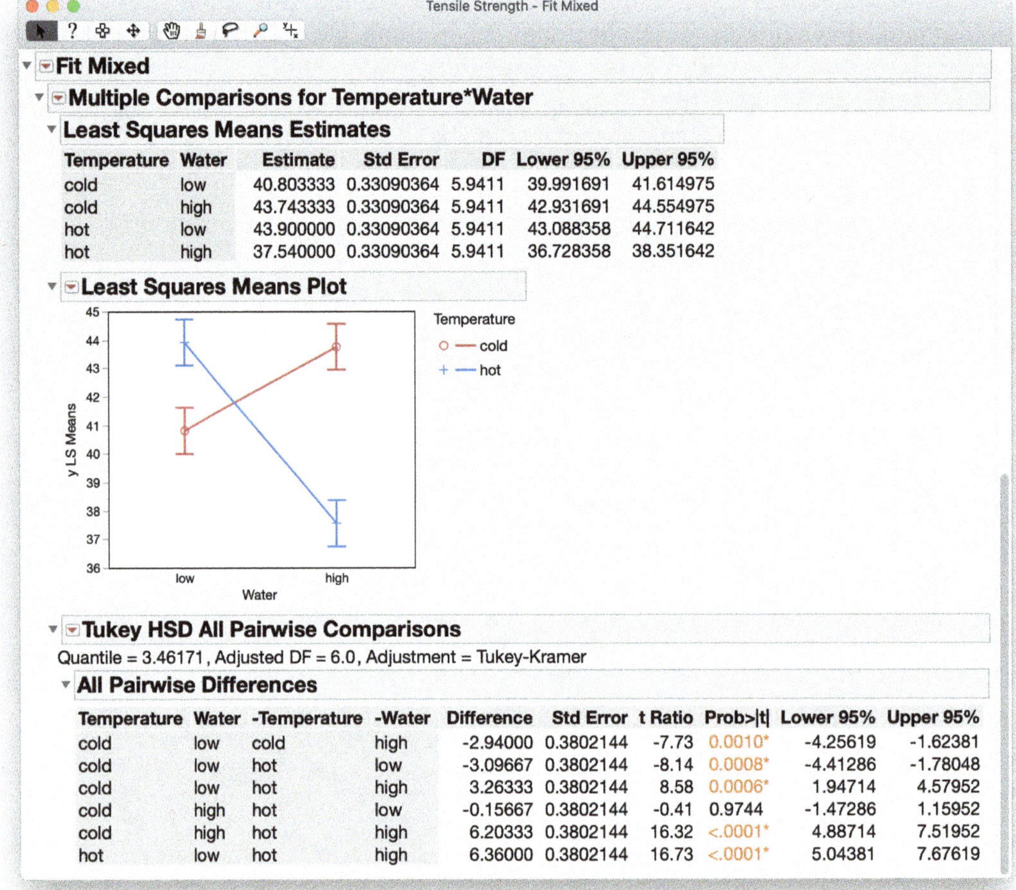

Tensile Strength - Fit Mixed

Fit Mixed

Multiple Comparisons for Temperature*Water

Least Squares Means Estimates

Temperature	Water	Estimate	Std Error	DF	Lower 95%	Upper 95%
cold	low	40.803333	0.33090364	5.9411	39.991691	41.614975
cold	high	43.743333	0.33090364	5.9411	42.931691	44.554975
hot	low	43.900000	0.33090364	5.9411	43.088358	44.711642
hot	high	37.540000	0.33090364	5.9411	36.728358	38.351642

Least Squares Means Plot

Tukey HSD All Pairwise Comparisons

Quantile = 3.46171, Adjusted DF = 6.0, Adjustment = Tukey-Kramer

All Pairwise Differences

Temperature	Water	-Temperature	-Water	Difference	Std Error	t Ratio	Prob>|t|	Lower 95%	Upper 95%
cold	low	cold	high	-2.94000	0.3802144	-7.73	0.0010*	-4.25619	-1.62381
cold	low	hot	low	-3.09667	0.3802144	-8.14	0.0008*	-4.41286	-1.78048
cold	low	hot	high	3.26333	0.3802144	8.58	0.0006*	1.94714	4.57952
cold	high	hot	low	-0.15667	0.3802144	-0.41	0.9744	-1.47286	1.15952
cold	high	hot	high	6.20333	0.3802144	16.32	<.0001*	4.88714	7.51952
hot	low	hot	high	6.36000	0.3802144	16.73	<.0001*	5.04381	7.67619

Again, due to the presence of the significant interaction, only four of the six comparisons presented are considered valid. One factor level must be held constant across the comparison. The differences, standard errors and p-values for the four valid comparisons match those obtained in the Slices option in Standard Least Squares.

3.4 Split-Plot Design, Greenhouse Example

A plant researcher has two plant varieties and a pesticide meant to protect the plants against disease. In the greenhouse the amount of pesticide, `Dose`, can be applied to sections of the benches. Due to procedure and equipment, this is a hard-to-change factor, or whole plot factor, in this experiment. The bench sections can hold multiple plants, so the two varieties of plant, `Type`, are randomly assigned spaces within each bench section. This is an easy-to-change factor, the split-plot factor. Due to the natural variability within the greenhouse, whole benches serve as the blocking criterion, `Block`. Figure 3.7 shows an example randomization of the two treatment factors to one bench. The researcher wants to identify the best plant and pesticide combination for disease resistance. The data for this experiment are in the `Variety-Pesticide Evaluation.jmp` data table.

Figure 3.7: Example Bench with Pesticide Dose and Variety Type Randomized by a Split-Plot Design

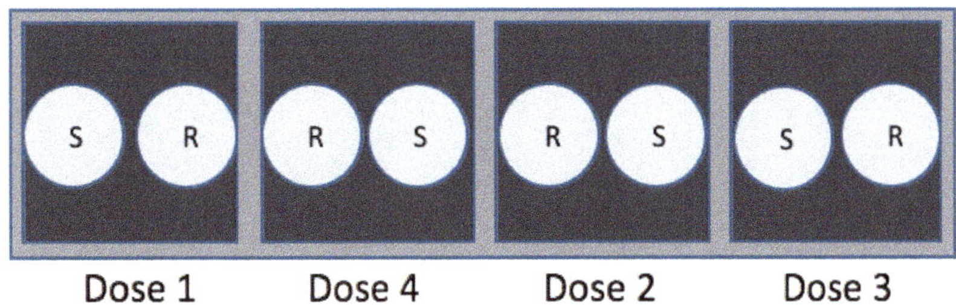

Given the description of the experiment and the sketched example bench, we can create the skeleton ANOVA of this design, which will lead us to our model. For the experiment design, there are five benches serving as blocks, four whole plots within each block (twenty total whole plots), and two split-plots within each whole plot. Dose (of pesticide) is applied to the whole plots, and Type (variety of plant) and Type*Dose are observed at the split-plot level.

Experiment Design		**Skeleton ANOVA**	
Source	***df***	**Source**	***df***
Block	5-1=4	Block	4
		Dose	4-1=3
WP(Block)	(4-1)*5=15	WP(Block)|Dose -> Block*Dose	12
		Type	(2-1)=1
		Type*Dose	3
SP(WP)	(2-1)*20=20	SP(WP)|Dose,Type -> Residual	20-4=16
Total	40-1=39	Total	39

Using the terms in the Skeleton ANOVA, we can define the statistical model for this split-plot experiment and use JMP to analyze it.

Mixed Model for a Split-Plot Experiment

The statistical model for these data is

$$y_{ijk} = \mu + r_k + \delta_i + w_{ik} + \tau_j + \delta\tau_{ij} + e_{ijk}$$

y_{ijk} is the observation on the ith `Dose`, j^{th} `Type`, and k^{th} block.
μ is the intercept.
r_k is the kth block effect, and $r_k \sim N(0, \sigma_r^2)$.
δ_i is the ith Dose effect.
w_{ik} is the ik^{th} whole-plot (Block * Dose) effect, and $w_{ik} \sim N(0, \sigma_w^2)$.
τ_j is the j_{th} Type effect.
$\delta\tau_{ij}$ is the ij^{th} Dose*Type interaction effect.
e_{ijk} is the ijk^{th} split-plot error and $e_{ijk} \sim N(0, \sigma^2)$.

JMP Instructions for Greenhouse Example

The Y column is the response variable *Y*. Add the model effects in the order of the statistical model shown above. Select the two columns `Type` and `Dose`, then use the *Macros > Full Factorial* macro. Next add the effect of `Block`. Select `Block` in the *Construct Model Effects* window then select `Dose` in the Columns list and click *Cross* to define the whole plot error term. (If we know the Block and the Dose level, we can uniquely identify which whole plot the observation is in.) Finally, select Block and Block*Dose and choose *Attributes > Random Effect* to designate them as random. The figure below shows the completed dialog box.

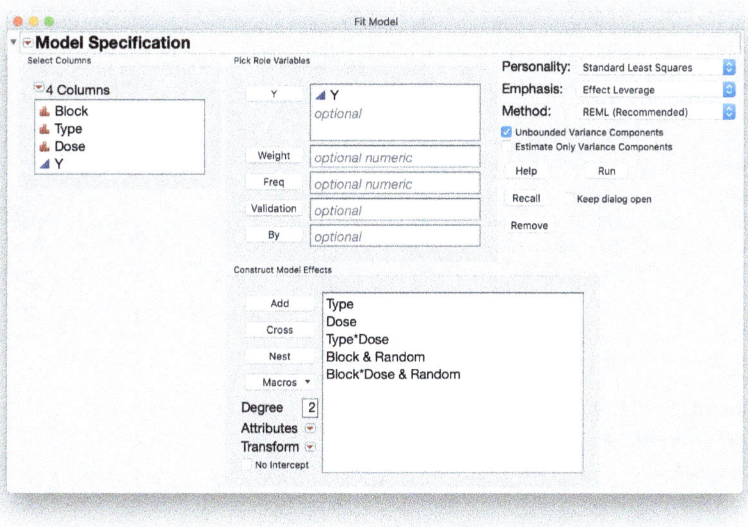

Results and Interpretation

Figure 3.8 shows the results of this analysis.

The Effect Summary section provides a quick visual indication of results for the fixed effects, indicating the primary action involves the main effect Dose. The Summary of Fit section provides basic model fitting statistics, and the Parameter Estimates section provides a breakdown of each individual fixed effects parameter.

The REML Variance Components section reveals that in partitioning of whole plot variance, bench-to-bench variability (Block) is approximately half of that within-bench (Block*Dose). Split-plot variability (Residual) is about the same as within-bench (Block*Dose). We therefore have approximately a 20:40:40 split attributable to these three sources of variability. The greenhouse manager may be interested in this breakdown and be able to offer further explanations in terms of the particular benches used

Figure 3.8: Results for the Greenhouse Example

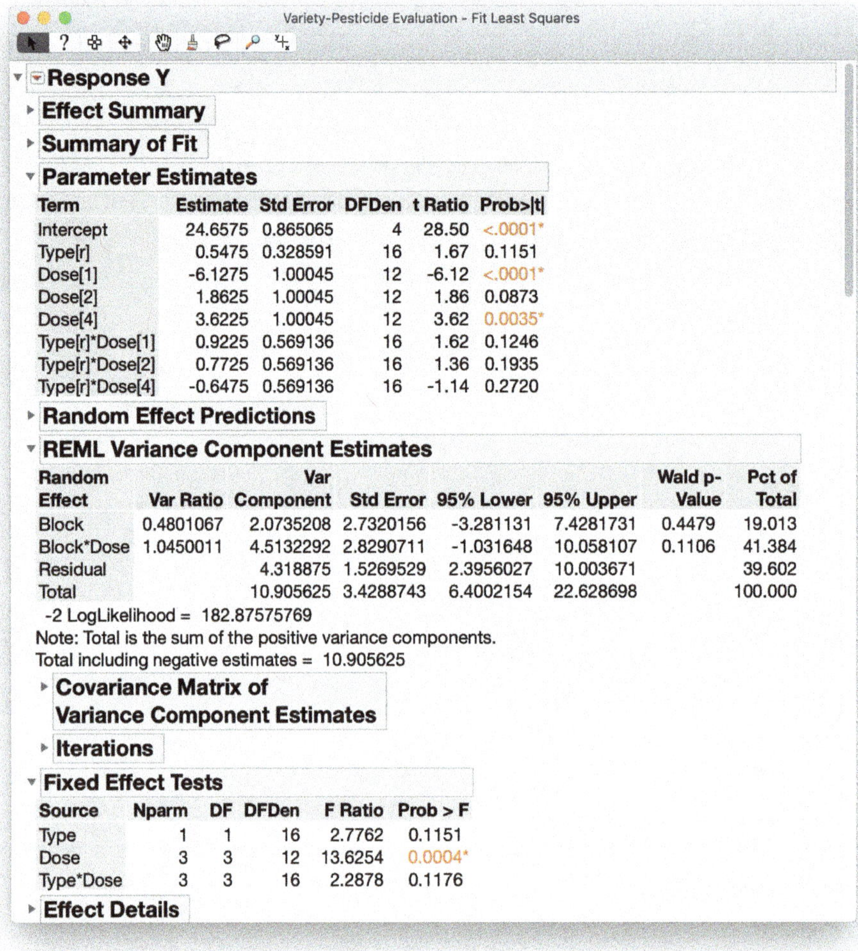

Variety-Pesticide Evaluation - Fit Least Squares

Response Y

Effect Summary

Summary of Fit

Parameter Estimates

Term	Estimate	Std Error	DFDen	t Ratio	Prob>|t|
Intercept	24.6575	0.865065	4	28.50	<.0001*
Type[r]	0.5475	0.328591	16	1.67	0.1151
Dose[1]	-6.1275	1.00045	12	-6.12	<.0001*
Dose[2]	1.8625	1.00045	12	1.86	0.0873
Dose[4]	3.6225	1.00045	12	3.62	0.0035*
Type[r]*Dose[1]	0.9225	0.569136	16	1.62	0.1246
Type[r]*Dose[2]	0.7725	0.569136	16	1.36	0.1935
Type[r]*Dose[4]	-0.6475	0.569136	16	-1.14	0.2720

Random Effect Predictions

REML Variance Component Estimates

Random Effect	Var Ratio	Var Component	Std Error	95% Lower	95% Upper	Wald p-Value	Pct of Total
Block	0.4801067	2.0735208	2.7320156	-3.281131	7.4281731	0.4479	19.013
Block*Dose	1.0450011	4.5132292	2.8290711	-1.031648	10.058107	0.1106	41.384
Residual		4.318875	1.5269529	2.3956027	10.003671		39.602
Total		10.905625	3.4288743	6.4002154	22.628698		100.000

-2 LogLikelihood = 182.87575769

Note: Total is the sum of the positive variance components.

Total including negative estimates = 10.905625

Covariance Matrix of Variance Component Estimates

Iterations

Fixed Effect Tests

Source	Nparm	DF	DFDen	F Ratio	Prob > F
Type	1	1	16	2.7762	0.1151
Dose	3	3	12	13.6254	0.0004*
Type*Dose	3	3	16	2.2878	0.1176

Effect Details

and potential differences in delivery of resources like sunlight and irrigation.

The interaction between plant `Type` and pesticide `Dose` appears to be negligible, though with multiple numerator degrees of freedom there may be an effect. Some researchers use a more lenient α-level, up to $\alpha = 0.20$, in these situations to avoid the Type I error of concluding there is no interaction when, in fact, there is. We can investigate this further using the Multiple Comparisons tool under Estimates in the red triangle menu.

Multiple Comparisons of Type*Dose Effect

Choose `Type*Dose` in the Effects list of the Multiple Comparison window. Check the *Show Least Squares Means Plot* box. Then check the *Create an Interaction Plot* box and choose `Dose` for the overlay. The figure to the right shows the completed Multiple Comparisons window for the interaction plot.

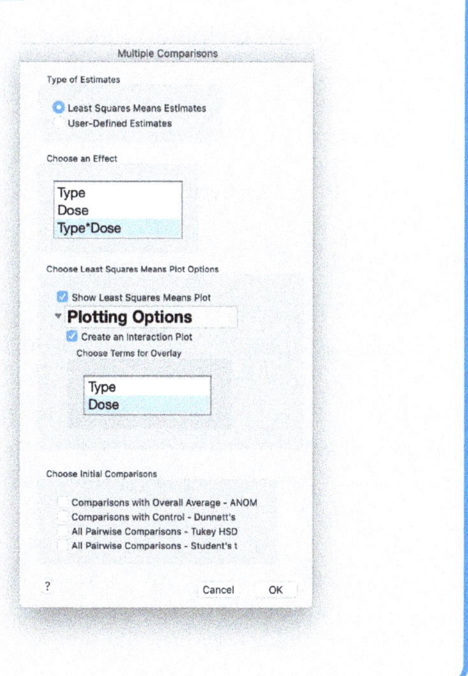

The Least Squares Means Plot defaults to including confidence limits for the means. They can be turned off in the Least Squares Means Plot red triangle menu. Figure 3.9 shows the Multiple Comparisons report with the limits turned off. Visually, it appears that plant Type S has a lower response from plant Type R with Doses 1 and 2, but similar or higher responses with Doses 4 and 8. Combined with the p-value for the interaction test, it may be better to draw conclusions based on the interaction effects rather than the main effects.

To formally test differences of the interaction effects, we can choose *All Pairwise Differences > Tukey HSD* from the Multiple Comparisons red triangle menu. This adds every pairwise difference of the Type*Dose Least Squares Means. However, not all of these comparisons are valid in the presence of an interaction effect. Only comparisons where either the `Type` is held constant while `Dose` changes, or `Dose` is held constant while `Type` changes are valid. These are the simple effects. Based on the interaction plot, we focus first on the simple effects of `Type` given a `Dose` level.

For Dose 1, the difference between the resistant, r, and susceptible, s, varieties is 2.94, which is the largest difference between the Types at any Dose level. This is not a significant difference, however. You can conclude that regardless of Dose level, there is no difference in the response of the two Types.

Looking instead at the effect of Dose given a specified Type, we find some interesting effects. Given the resistant Type, Dose 1 results in a significantly lower response than Doses 2 and 4 (p-values 0.0128 and 0.0090, respectively) But there is no evidence of a statistical difference between Doses 1 and 8, 2 and 4, 2 and 8, or 4 and 8 (all p-values > 0.6298). Given the resistant Type, there are diminishing returns when increasing the pesticide Dose.

When holding Type constant at the susceptible variety, Dose 1 is significantly lower than all three other Dose levels (p-values 0.0094, 0.0004, and 0.0050, respectively). Like the resistant Type, the higher dose comparisons show no difference.

The alternate conclusion about the difference between Doses 1 and 8 with the two variety Types explains why the test for interaction was marginal. Whether that difference is important would be a decision for the plant researchers. However, given that Doses 2, 4, and 8 are not significantly different for either Type, it may be sufficient to conclude that a higher Dose of pesticide regardless of susceptibility of plant Type is required. That is what an analysis of the main effect of Dose would also conclude.

Figure 3.9: Multiple Comparisons for Type*Dose

Variety-Pesticide Evaluation - Fit Least Squares

Response Y

Multiple Comparisons for Type*Dose

Least Squares Means Plot

Dose
1
2
4
8

Y LS Means
30
25
20
15
r
s
Type

Tukey HSD All Pairwise Comparisons

Quantile = 3.46215, Adjusted DF = 16.0, Adjustment = Tukey-Kramer

All Pairwise Differences

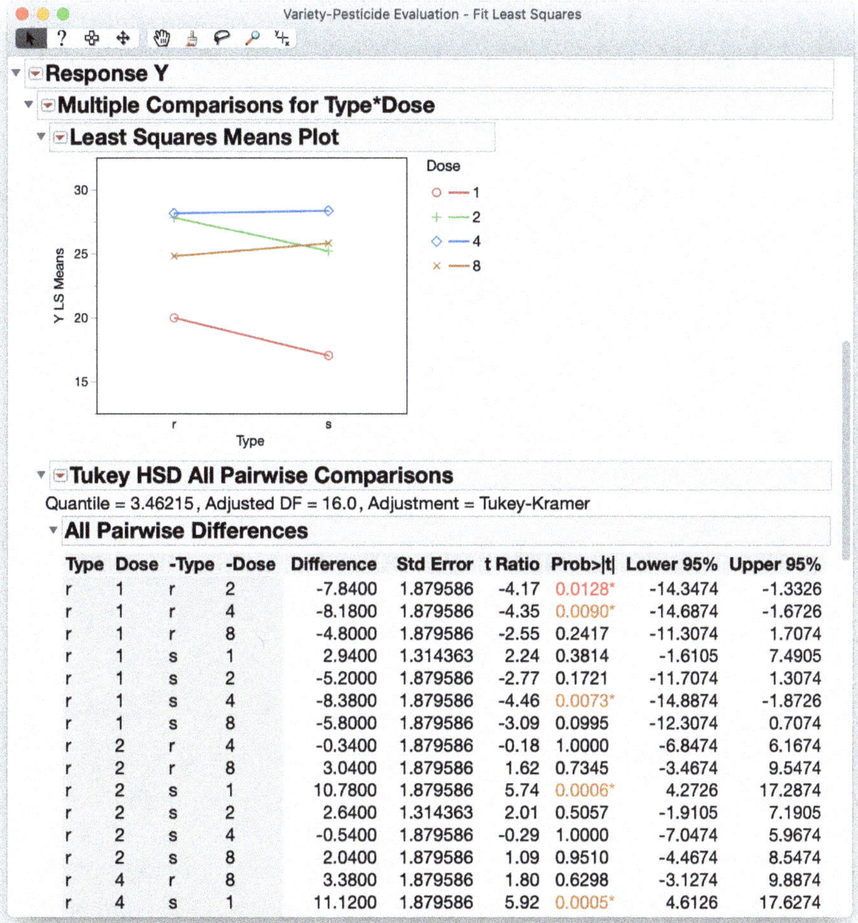

Type	Dose	-Type	-Dose	Difference	Std Error	t Ratio	Prob>|t|	Lower 95%	Upper 95%
r	1	r	2	-7.8400	1.879586	-4.17	0.0128*	-14.3474	-1.3326
r	1	r	4	-8.1800	1.879586	-4.35	0.0090*	-14.6874	-1.6726
r	1	r	8	-4.8000	1.879586	-2.55	0.2417	-11.3074	1.7074
r	1	s	1	2.9400	1.314363	2.24	0.3814	-1.6105	7.4905
r	1	s	2	-5.2000	1.879586	-2.77	0.1721	-11.7074	1.3074
r	1	s	4	-8.3800	1.879586	-4.46	0.0073*	-14.8874	-1.8726
r	1	s	8	-5.8000	1.879586	-3.09	0.0995	-12.3074	0.7074
r	2	r	4	-0.3400	1.879586	-0.18	1.0000	-6.8474	6.1674
r	2	r	8	3.0400	1.879586	1.62	0.7345	-3.4674	9.5474
r	2	s	1	10.7800	1.879586	5.74	0.0006*	4.2726	17.2874
r	2	s	2	2.6400	1.314363	2.01	0.5057	-1.9105	7.1905
r	2	s	4	-0.5400	1.879586	-0.29	1.0000	-7.0474	5.9674
r	2	s	8	2.0400	1.879586	1.09	0.9510	-4.4674	8.5474
r	4	r	8	3.3800	1.879586	1.80	0.6298	-3.1274	9.8874
r	4	s	1	11.1200	1.879586	5.92	0.0005*	4.6126	17.6274

JMP Pro Instructions for Greenhouse Example

The `Y` column is the response variable Y. Add the model effects in the order of the statistical model. For the Fixed Effects select the two columns `Type` and `Dose`, then use the *Macros > Full Factorial* macro. Switch to the Random Effects tab to add the remaining effects from the model. Add the effect of `Block`. Select `Block` in the Random Effects window then select `Dose` in the Columns list and click *Cross* to define the whole plot error term. If we know the `Block` and the `Dose` level, we can uniquely identify which whole plot the observation is in. The completed dialog box is shown below.

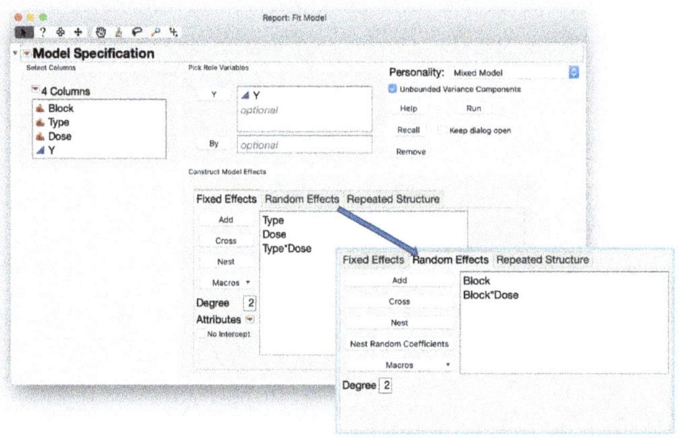

Results and Interpretation

The results of the Mixed Model personality analysis are shown in Figure 3.10 and are the same as those shown in the previous section using the Standard Least Squares personality. The Type*Dose interaction is negligible, and the main effect of `Type` is also negligible at usual α-levels. We do have a significant main effect of `Dose` with p-value of 0.0004. We can investigate this effect further using *Multiple Comparisons* from the Fit Mixed red triangle menu. In the interest of space, the steps are not shown, but the results are included in Figure 3.11. Dose 1 results in a significantly lower response than the other three Doses (p-values 0.0018, 0.0003, and 0.0064, respectively). Those three Doses are not significantly different from each other, however, with all p-values > 0.3095. We can conclude that a Dose greater than Dose 1 is needed to improve the response of the plants, but more than Dose 2 may be unnecessary and potentially wasteful of the product and harmful to the environment.

Figure 3.10: Fit Mixed Report for the Variety-Pesticide Data

Variety-Pesticide Evaluation - Fit Mixed

Fit Mixed

Actual by Predicted Plot

Actual by Conditional Predicted Plot

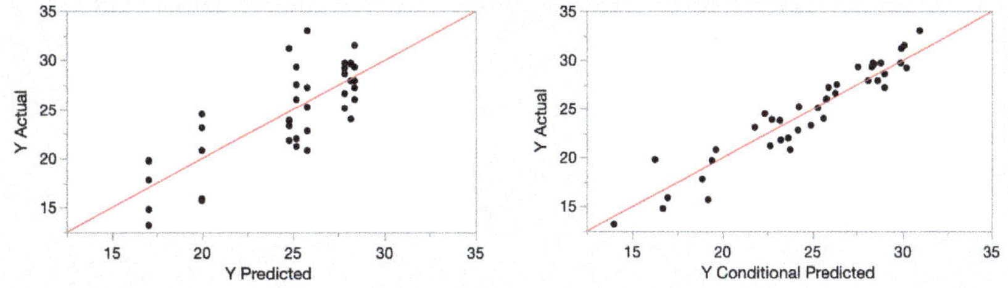

Fit Statistics

-2 Residual Log Likelihood	182.87576
-2 Log Likelihood	190.63737
AICc	222.06595
BIC	231.21505

Random Effects Covariance Parameter Estimates

Variance Component	Estimate	Std Error	95% Lower	95% Upper	Wald p-Value
Block	2.0735208	2.7320156	-3.281131	7.4281731	0.4479
Block*Dose	4.5132292	2.8290711	-1.031648	10.058107	0.1106
Residual	4.318875	1.5269529	2.3956027	10.003671	
Total	10.905625	3.4288743	6.4002154	22.628698	

Fixed Effects Parameter Estimates

Term	Estimate	Std Error	DFDen	t Ratio	Prob>|t|	95% Lower	95% Upper
Intercept	24.6575	0.865065	4.0	28.50	<.0001*	22.255694	27.059306
Type[r]	0.5475	0.3285907	16.0	1.67	0.1151	-0.149081	1.2440813
Dose[1]	-6.1275	1.0004499	12.0	-6.12	<.0001*	-8.307293	-3.947707
Dose[2]	1.8625	1.0004499	12.0	1.86	0.0873	-0.317293	4.0422931
Dose[4]	3.6225	1.0004499	12.0	3.62	0.0035*	1.4427069	5.8022931
Type[r]*Dose[1]	0.9225	0.5691359	16.0	1.62	0.1246	-0.284014	2.1290141
Type[r]*Dose[2]	0.7725	0.5691359	16.0	1.36	0.1935	-0.434014	1.9790141
Type[r]*Dose[4]	-0.6475	0.5691359	16.0	-1.14	0.2720	-1.854014	0.5590141

Figure 3.11: Fit Mixed Report for the Variety-Pesticide Data - Pairwise Differences

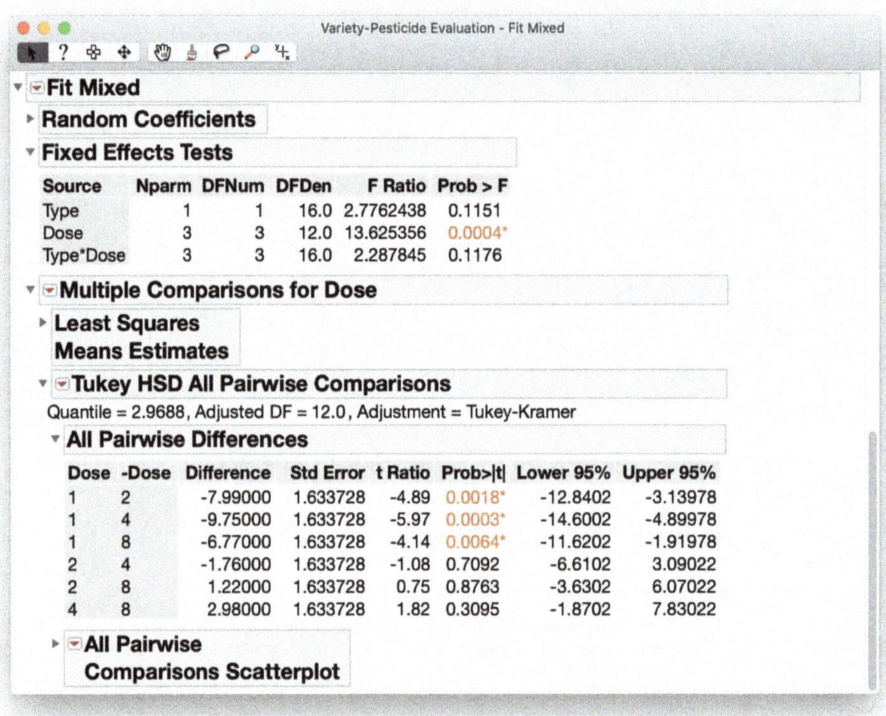

Variety-Pesticide Evaluation - Fit Mixed

Fit Mixed

Random Coefficients

Fixed Effects Tests

Source	Nparm	DFNum	DFDen	F Ratio	Prob > F
Type	1	1	16.0	2.7762438	0.1151
Dose	3	3	12.0	13.625356	0.0004*
Type*Dose	3	3	16.0	2.287845	0.1176

Multiple Comparisons for Dose

Least Squares Means Estimates

Tukey HSD All Pairwise Comparisons

Quantile = 2.9688, Adjusted DF = 12.0, Adjustment = Tukey-Kramer

All Pairwise Differences

Dose	-Dose	Difference	Std Error	t Ratio	Prob>|t|	Lower 95%	Upper 95%
1	2	-7.99000	1.633728	-4.89	0.0018*	-12.8402	-3.13978
1	4	-9.75000	1.633728	-5.97	0.0003*	-14.6002	-4.89978
1	8	-6.77000	1.633728	-4.14	0.0064*	-11.6202	-1.91978
2	4	-1.76000	1.633728	-1.08	0.7092	-6.6102	3.09022
2	8	1.22000	1.633728	0.75	0.8763	-3.6302	6.07022
4	8	2.98000	1.633728	1.82	0.3095	-1.8702	7.83022

All Pairwise Comparisons Scatterplot

3.5 What About Interactions Between Fixed and Random Effects?

Generally, the interaction between a fixed effect and a random effect will not be of interest for statistical testing, but can be effective for modeling a source of variability. Be aware though, in cases such as the metal bonding example from Chapter 2, this interaction effect is completely confounded with residual errors, and should therefore not even be included in the model. On the other hand, if the treatment*block combinations are replicated, then it can be distinguished from residual variance and we can describe things about that interaction effect. But how? As a fixed effect? Or as a random effect?

Here we see an RCBD with no replication, where the interaction combinations are the smallest unit that receives the treatment application. We could name the block level and the treatment level to completely identify any experimental unit. Notice that this also means that we have the same number of interaction combinations as we have experimental units for the whole experiment.

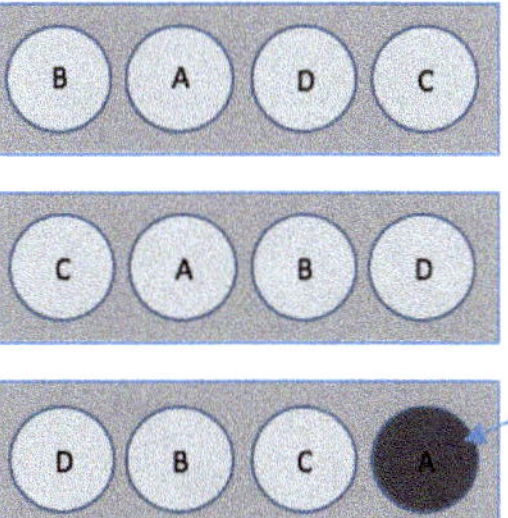

Looking more closely at this design, we see that there are three replicates, each in different blocks, that will lend themselves to an estimate and a test of the Treatment A effect.

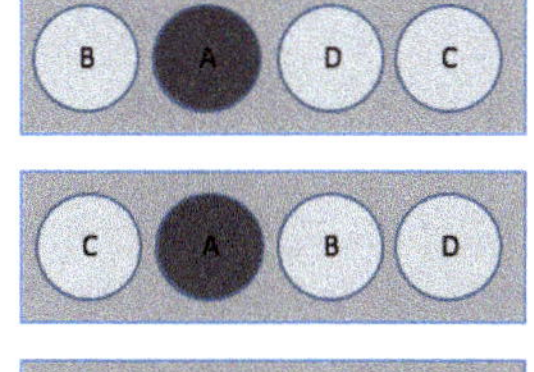

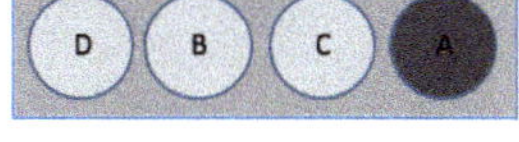

We see that there are four replicates, all in a single block but given the four different treatments, that allow us to make an estimate and get a statistical test of the Block 3 effect.

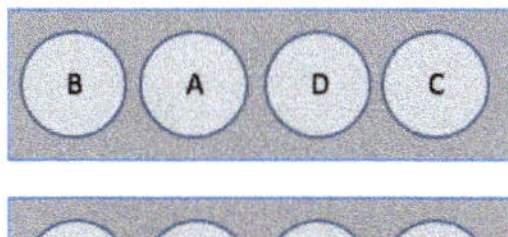

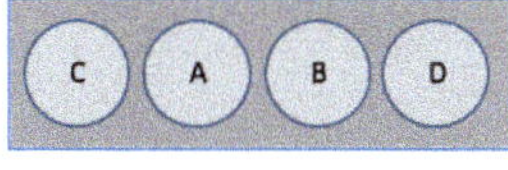

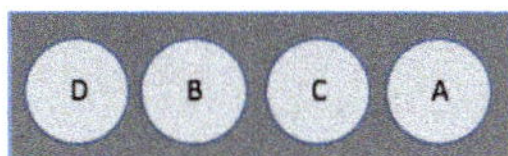

But when we have no replication for the interaction effect between the blocks and the treatment, we are not able to get a statistical test for that effect. We need at least two observations at the same factor combination in order to get an estimate of the corresponding variance. Without a variance estimate, we cannot do traditional statistical testing.

Here we see the same experimental and treatment design, but with replication at the block*treatment level.

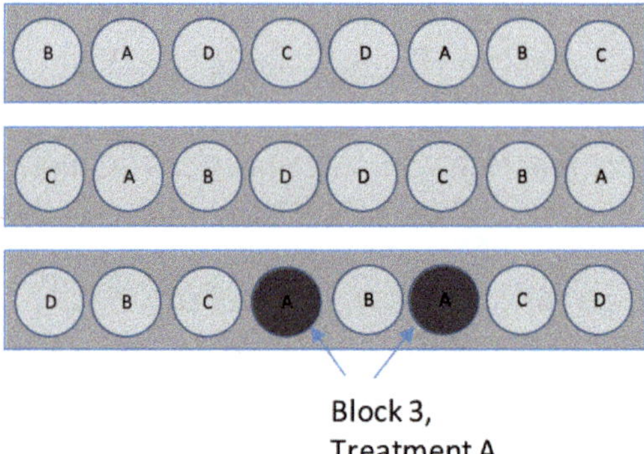

Now we have twice as many experimental units as we have treatment*block combinations, and just specifying a certain treatment*block combination identifies *two* experimental units instead of a single experimental unit. We now have replication at the interaction level, so we can now estimate that interaction effect if we want to do so.

The next question, though, is "Do we *want* to estimate that interaction effect, even if we have the replication that allows it?"

Remember that a conceptual difference between random and fixed effects is that we consider the levels of the random effect to be (at least approximately) randomly sampled from a larger set of possible levels, so it is usually not interesting to test differences between these random levels. The same is true for the interactions between a random and a fixed effect. If you did not care about making comparisons between the blocks in an RCBD, then you likely also do not care about making comparisons between the block-by-treatment combinations. By treating the interaction of a fixed effect with a random effect as random, the focus of that effect is on the contribution to explaining variation, rather than on making means comparisons as we would for a fixed-by-fixed interaction.

It can be helpful to think about the *randomness* attribute of a factor as a contagious disease – if a factor is random, then any interactions that include that factor will also be random. A typical mixed model would be as follows:

Statistical Model: One Fixed , One Random, and Their (Random) Interaction

The statistical model with replication inside the block*treatment interaction and with the block*treatment interaction included in the model is

$$y_{ijk} = \mu + \alpha_i + r_j + \alpha r_{ij} + e_{ijk}$$

y_{ijk} is the continuous response variable.
α_i is the effect of the ith level of the treatment.
r_j is the effect of the jth level of the random (blocking) factor and $r_j \sim N(0,\sigma_r^2)$.
αr_{ij} is the effect of the ith level of the treatment at the jth level of the random (blocking) factor and $\alpha r_{ij} \sim N(0,\sigma_{ar}^2)$.
e_{ijk} is the residual error and $e_{ijk} \sim N(0,\sigma^2)$.

This model requires replication within the interaction. That is, there must be multiple observations for any single combination of the ith level of the treatment at the rth level of the random (blocking) factor.

One final note about interactions between fixed and random effects: we often find that experimental units are different for different treatment factors within the same experiment. When deciding whether to include a mixed interaction term, think about the experimental unit for each treatment separately. You still might include a mixed interaction effect that is equivalent to the experimental unit for one treatment factor but has replication for another treatment factor. In fact, this is exactly what we do for split-plot-style experiments. To illustrate this further, let's now consider another form of split-plot experiment that uses nesting.

3.6 Nested Design, Semiconductor Example

The `Semiconductor Experiment.jmp` data table contains the results of the semiconductor experiment described at the beginning of this chapter. The column **`Wafer`** contains identifiers for the twelve randomly selected wafers from a lot. Importantly, note how Wafer is coded in the table. Wafer only includes the numbers 1, 2, and 3, and those values repeat for each of the process conditions, `ET`. However, Wafer 1 with ET 1 is not the same as Wafer 1 with ET 2. Therefore, Wafer is nested within levels of ET and is the experimental unit for ET. `Resistance` is measured at four `Position` locations on each Wafer. The ANOVA is as follows:

Experiment Design		Skeleton ANOVA	
Source	***df***	**Source**	***df***
		ET	4-1=3
Wafer	(12-1)=11	Wafer|ET -> Wafer(ET)	11-6=8
		Position	(4-1)=3
		ET*Position	9
Quadrant(Wafer)	(4-1)*12=36	Quad(Wafer)|ET,Pos. -> Residual	36-12=24
Total	48-1=47	Total	47

This is a two-way factorial treatment design with factors ET and Position, but the experimental units are different for each. We label this example "Nested Design" to emphasize how Wafer is coded, but it is also a split-plot design with ET as the whole-plot (hard-to-change) factor and Position as the subplot (easy-to-change) factor. Wafer denotes the whole-plot experimental units for ET, and Quadrant(Wafer) denotes the subplot experimental units for Position and ET*Position.

Mixed Model for the Nested Design, Semiconductor Example

The statistical model for these data is

$$y_{ijk} = \mu + \alpha_i + w_{ij} + \beta_k + (\alpha\beta)_{ik} + e_{ijk}$$

y_{ijk} is the `Resistance` measured at the kth `Position` on the jth `Wafer` in the ith level of `ET`.
α_i is the effect of the ith level of `ET`.
w_{ij} is the effect of the jth `Wafer` in the ith `ET`, and $w_{ij} \sim N(0,\sigma_w^2)$.
β_k is the effect of the kth level of `Position`.
$(\alpha\beta)_{ik}$ is the interaction between the ith level of `ET` and k^{th} `Position`.
e_{ijk} is the residual error and $e_{ijk} \sim N(0,\sigma^2)$.

JMP Instructions for Semiconductor Experiment.jmp example

Below is the completed Fit Model dialog box. **Resistance** is the *Y* variable. **ET**, **Wafer** and **Position** are the Model Effects. Because Wafer is nested within ET, select Wafer in the Model Effects and select ET in the Select Columns section, and then click the *Nest* button. Wafer in the Model Effects is now Wafer[ET]. This is the standard notation for a nested effect: we say "Wafer nested in ET" and we denote this with InsideEffect[OutsideEffect] (see Tip Box, below). This effect is also the whole plot error term and must be random. With it selected, choose *Attributes* > *Random Effect* to designate it as random. To add the interaction term for ET and Position, select **Position** in the Model Effects and select **ET** in Select Columns, then click the *Cross* button. The effect Position*ET is added to the Model Effects list. Click the *Run* button to fit the model.

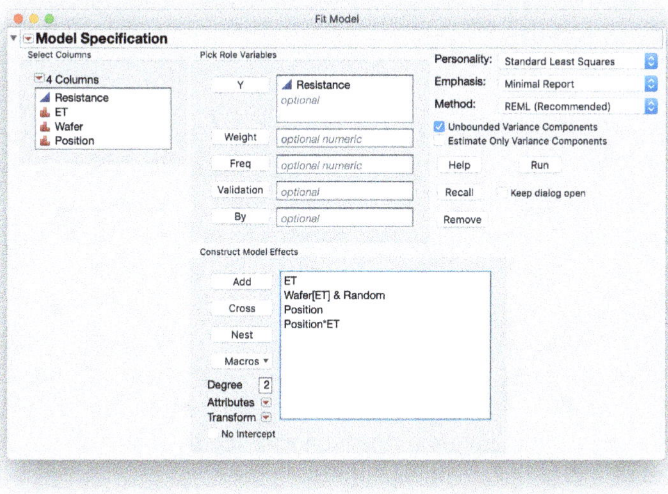

Tip Box

The notation of nested effects is often a point of confusion for new Mixed Modelers. We always say "Effect A is nested in Effect B" and write A[B]. If we took a sample of randomly selected potato chip samples from different supplier plants, we would say that the potato chip samples are nested in the supplier plant, and we would write this PotatoChipSample[SupplierPlant]. We go first to the Supplier Plant, and then we sample within it to get the Potato Chips Samples, so some people feel that the order of the notation is "backwards." It might help to think of the logic this way: you are writing down the effect, A, (or "PotatoChipSamples"), and you want to give yourself a parenthetical reminder that it is actually nested inside another Effect, B, (or "SupplierPlant"), hence A[B], or PotatoChipSample[SupplierPlant].

Results and Interpretation

The analysis of these data is shown in Figure 3.12.

In the REML Variance Component Estimates table you see that the variability among Wafers (the Wafer[ET] Variance Component) is approximately the same as residual variability within Wafer (the Residual Variance Component), each contributing half of the total variability in the data after accounting for fixed effects. The Fixed Effects Tests table shows that there is not a significant interaction between Position and ET ($p = 0.6125$), nor is there a significant main effect of ET ($p = 0.2015$). You can see that the denominator degrees of freedom for the test of ET is 8, as it is properly using the Wafer[ET] whole plot variance for its error term. The lack of statistical significance is evidence for a small effect that likely is not strictly zero. If the engineers at the semiconductor plant believe ET truly affects Resistance, the variance estimates from this experiment can be used to conduct a power analysis to determine the number of wafers necessary to detect a significant difference with a predefined magnitude.

Figure 3.12: Standard Least Squares Report for the `Semiconductor Experiment` Data

Semiconductor Experiment - Fit Least Squares

Response Resistance

Effect Summary

Source	LogWorth	PValue
Position	1.462	0.03451
ET	0.696	0.20150
Position*ET	0.213	0.61253

Remove Add Edit FDR

Summary of Fit

RSquare	0.76214
RSquare Adj	0.650643
Root Mean Square Error	0.333391
Mean of Response	6.002917
Observations (or Sum Wgts)	48

Parameter Estimates

Random Effect Predictions

REML Variance Component Estimates

Random Effect	Var Ratio	Var Component	Std Error	95% Lower	95% Upper	Wald p-Value	Pct of Total
Wafer[ET]	0.9517853	0.1057903	0.0672688	-0.026054	0.2376347	0.1158	48.765
Residual		0.1111493	0.032086	0.0677669	0.2151077		51.235
Total		0.2169396	0.0709919	0.1249573	0.466412		100.000

-2 LogLikelihood = 72.831204814

Note: Total is the sum of the positive variance components.

Total including negative estimates = 0.2169396

Covariance Matrix of Variance Component Estimates

Iterations

Fixed Effect Tests

Source	Nparm	DF	DFDen	F Ratio	Prob > F
ET	3	3	8	1.9415	0.2015
Position	3	3	24	3.3855	0.0345*
Position*ET	9	9	24	0.8092	0.6125

Effect Details

Similar considerations apply to the Position*ET interaction. One of the main questions the engineers have is whether there is any type of interaction between the two effects. We have the statistical answer to this question from this Fixed Effects Test output: the test for the Position*ET interaction is not statistically significant. However, we may still want to explore the magnitude of the effect, despite it not being statistically significant.

You can visualize the interaction plot from within the Multiple Comparisons tool. *Estimates > Multiple Comparisons* under the red triangle menu brings up the Multiple Comparisons dialog box. Select Position*ET, check the Show Least Squares Means Plot box, check the Create an Interaction Plot box, and choose Position as the overlay term. The completed dialog box is shown in Figure 3.13.

Figure 3.13: Multiple Comparisons Dialog Box for Interaction Plot

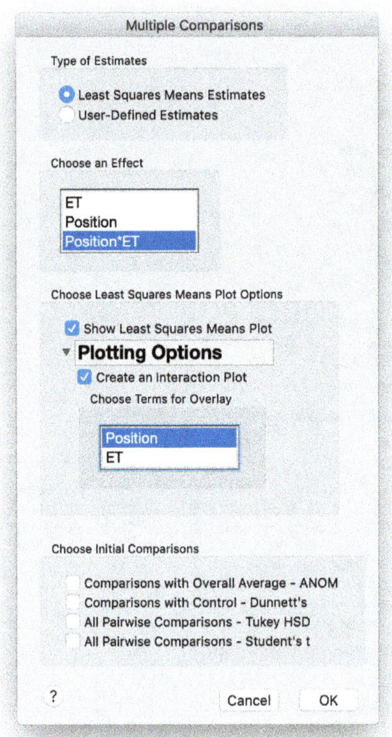

After clicking *OK*, the Multiple Comparisons for Position*ET section is added to the report. It includes a table of the Least Squares Means Estimates as well as the Least Squares Means Plot. The error bars on the plot can obscure the visualization of the interaction, so you can turn them off by selecting Show Confidence Limits from the Least Squares Means Plot red triangle menu. The plot without the confidence bars is shown in Figure 3.14.

Figure 3.14: Interaction Plot for the `Semiconductor Experiment` Data

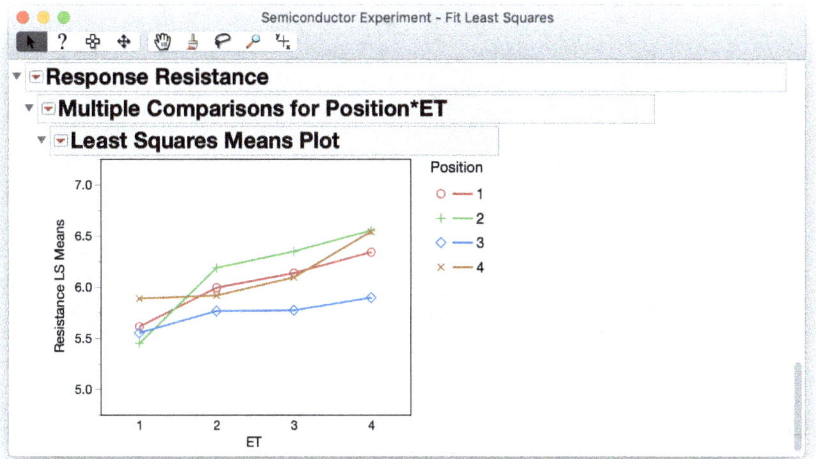

While the lines for Position cross a little from one level of ET to the next, it does appear that the effect of ET is the same regardless of Position; that is, the lines are approximately parallel to each other. This visually confirms the statistical test initially performed. The plot also shows that for all types of ET except 1, Position 3 has the lowest resistance measurement. You can confirm the difference is significant using Multiple Comparisons of just the Position effect. The dialog box for this comparison is shown in Figure 3.15.

Figure 3.15: Multiple Comparison Dialog for Least Squares Means of Position

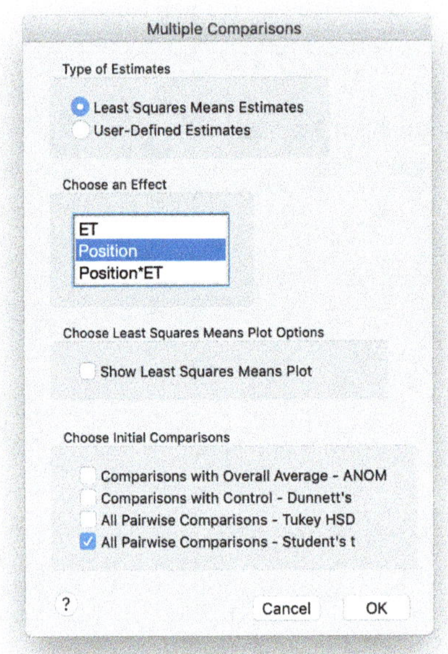

The results of these comparisons are shown in Figure 3.16. You can see in the All Pairwise Differences table that Position 3 has significantly less resistance than both Position 2 and Position 4. Position 3 would have significantly less resistance than Position 1 with a slightly relaxed α-level of $\alpha = 0.0560$.

In previous examples, we used the Tukey correction to control the familywise error rate, but we did not in this example. To use or not use the correction method is a choice the analyst must make weighing the design of the study, the purpose of the study, and the number of comparisons being made. If the design was powered for the type and number of comparisons made (that is, these comparisons were pre-planned and not post-hoc), typically no correction would be used as that works against the design. If a study is a preliminary, exploratory experiment where the researchers are looking for any possible treatment to move on to further study, adjustments would not be used as they make things "less significant" and a potential treatment might be overlooked. If making a Type 1 error would have negative consequences, using the correction would be recommended. For this example, despite the comparisons being post-hoc, we chose not to use the correction as the number of comparisons was small and the risks of a Type 1 error were negligible.

Figure 3.16: Least Squares Means Differences of Position

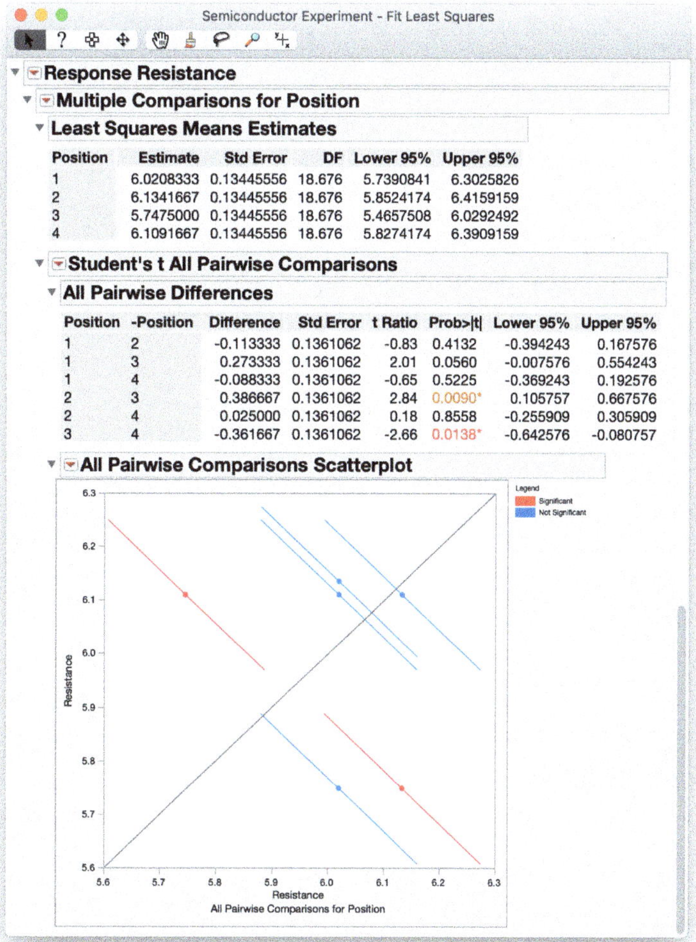

Semiconductor Experiment - Fit Least Squares

Response Resistance

Multiple Comparisons for Position

Least Squares Means Estimates

Position	Estimate	Std Error	DF	Lower 95%	Upper 95%
1	6.0208333	0.13445556	18.676	5.7390841	6.3025826
2	6.1341667	0.13445556	18.676	5.8524174	6.4159159
3	5.7475000	0.13445556	18.676	5.4657508	6.0292492
4	6.1091667	0.13445556	18.676	5.8274174	6.3909159

Student's t All Pairwise Comparisons

All Pairwise Differences

Position	-Position	Difference	Std Error	t Ratio	Prob>|t|	Lower 95%	Upper 95%
1	2	-0.113333	0.1361062	-0.83	0.4132	-0.394243	0.167576
1	3	0.273333	0.1361062	2.01	0.0560	-0.007576	0.554243
1	4	-0.088333	0.1361062	-0.65	0.5225	-0.369243	0.192576
2	3	0.386667	0.1361062	2.84	0.0090*	0.105757	0.667576
2	4	0.025000	0.1361062	0.18	0.8558	-0.255909	0.305909
3	4	-0.361667	0.1361062	-2.66	0.0138*	-0.642576	-0.080757

All Pairwise Comparisons Scatterplot

Legend
Significant
Not Significant

Resistance
All Pairwise Comparisons for Position

JMP Pro Instructions for Semiconductor Experiment.jmp Example

As this is a factorial experiment with the two factors ET and Position, you can use the *Macros > Full Factorial* macro when using the Mixed Model personality of Fit Model. That completes the Fixed Effects tab of the launch dialog box. The Random Effects tab gets the term identifying the whole plot variance, Wafer[ET]. Add `Wafer` to the Random Effects. Select it in the Random Effects window and select `ET` in the columns window, and then click the *Nest* button to nest the Wafers within ET. The completed dialog box is shown below.

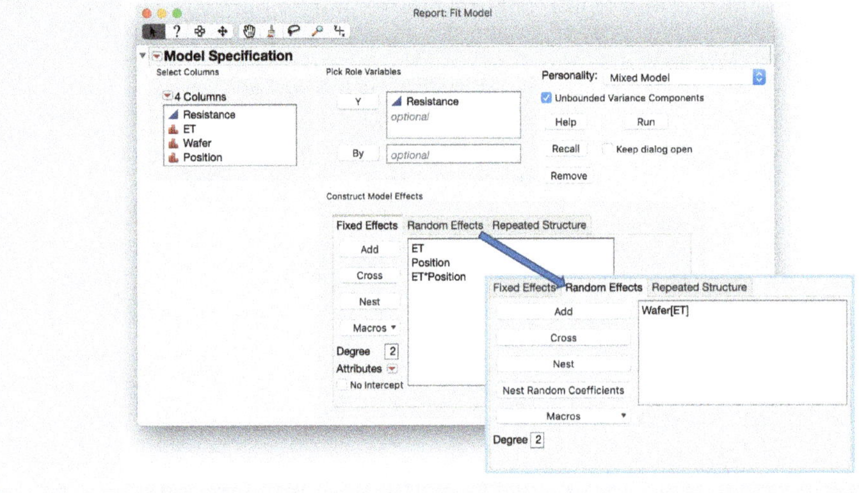

Results and Interpretation

Figure 3.17 shows the results of the analysis from the completed dialog box.

The Actual by Predicted Plot and Actual by Conditional Predicted Report indicate a reasonable fit for this model to these data. The Random Effects Covariance Parameter Estimates table shows the estimates and standard errors of the two variance components, the whole plot Wafer[ET] and residual variance, approximately equal. The Fixed Effects Tests table shows the same results as in Standard Least Squares with no significant effects of the interaction between ET and Position and a mildly significant main effect of Position.

The remainder of the analysis to answer the engineer's questions can be performed the same way as in the previous Standard Least Squares personality section.

Figure 3.17: Fit Mixed Report for the `Semiconductor Experiment` Data

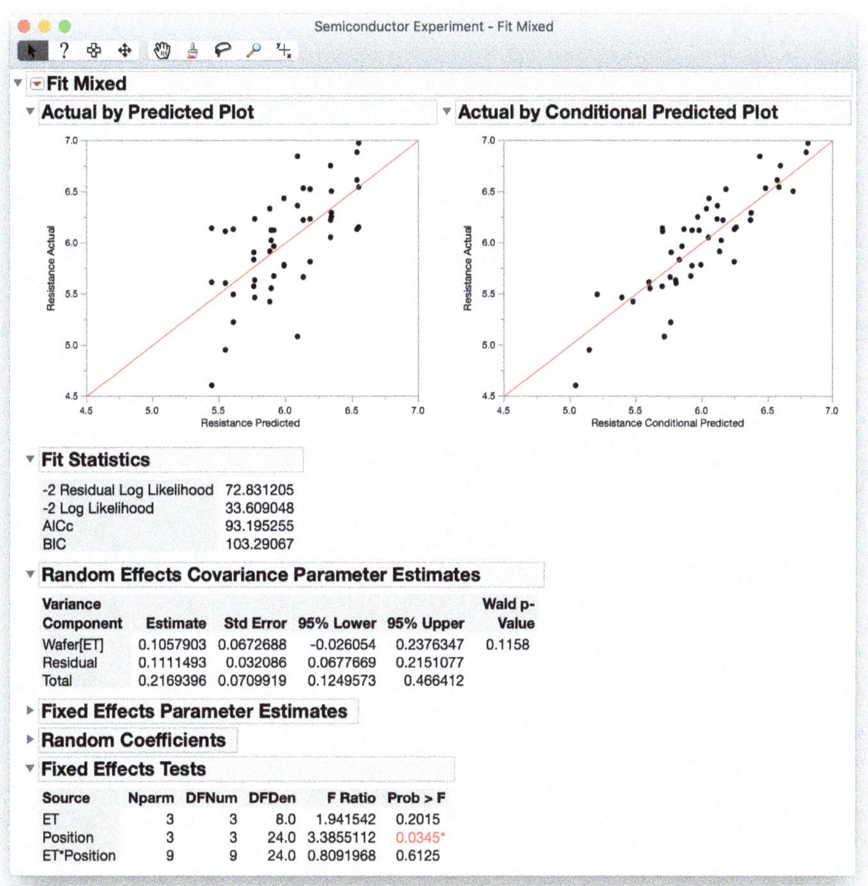

Fit Statistics	
-2 Residual Log Likelihood	72.831205
-2 Log Likelihood	33.609048
AICc	93.195255
BIC	103.29067

Variance Component	Estimate	Std Error	95% Lower	95% Upper	Wald p-Value
Wafer[ET]	0.1057903	0.0672688	-0.026054	0.2376347	0.1158
Residual	0.1111493	0.032086	0.0677669	0.2151077	
Total	0.2169396	0.0709919	0.1249573	0.466412	

Source	Nparm	DFNum	DFDen	F Ratio	Prob > F
ET	3	3	8.0	1.941542	0.2015
Position	3	3	24.0	3.3855112	0.0345*
ET*Position	9	9	24.0	0.8091968	0.6125

3.7 Exercises

1. In the tensile strength example, what are the experimental units to which treatments are applied? How does this compare to treatment assignment in a split-plot design?
2. In the semiconductor example, recode the levels of Wafer in the table so that they are no longer nested within ET, and rerun the analysis. What changes in the results?

Chapter 4

Multiple Random Effects

In this chapter, we investigate experiments with more than one random effect. There are many different ways this can occur, and an exhaustive list would be nearly impossible to compile. We will focus on a few of the most common. We will also explore the modeling of random effects a bit deeper, encountering an example with a negative estimated block variance.

4.1 Motivating Examples

Fabric Shrinkage — A fabric manufacturer needs to test four new materials to be used in permanent press garments. The heat chamber used to test fabrics has four positions. Each fabric should be tested under each position, and due to time constraints, the manufacturer is limited to four runs. Fabric shrinkage is the response of interest.

Mouse Condition — A trial involving laboratory mice investigates the effects of four housing conditions (CONDITION 1, 2, 3, and 4) and three feeding regimens (DIET "restricted," "normal," and "supplemental") on weight gain in mice. As described in Example 5.7 in Stroup et al. (2018), two housing conditions can be accommodated in one CAGE unit, and within each housing-condition-in-a-caging-unit, mice can be separated into the three distinct diet groups.

4.2 Conceptual Background

Models with multiple random effects arise in several ways. One way is when an experiment has more than one blocking criteria, such as runs and positions in the opening example. There is still only one treatment factor, Material, but its levels have been randomized in two directions (Run and Position) to balance variability. Run and Position correspond to rows and columns in a *Latin Square* design.

Multiple blocking factors can also occur in a *split-plot* manner, like we saw in Chapter 3 for the application of two treatment factors with one randomized within blocks that are defined by the other. These designs with two or more treatment factors such that one is harder to change the levels of than the other can also give rise to many variations of the split plot experiment design, with nested or independent subplots.

In the split-plot design, the hard-to-change factor is assigned to the larger experimental unit, known as the whole plot. This assignment can be done either in a completely randomized manner, or, when there is a need to block, in some form of a blocked design (RCBD, BIBD, etc.). Each of the whole plot units are then subdivided into split-plot units to which the easy-to-change factor is assigned. When the split-plot design is randomized within blocks, we will have two random effects: the high-level blocks and the whole-plot units.

Random effects might also arise from a *multilevel* or *nested* model. The classic example is assessing instructional methods by sampling CLASSROOMS from within SCHOOLS from within DISTRICTS. The two levels of experimental units in a split-plot design can also be viewed as a multilevel model. Multilevel models are covered in Chapter 5.

4.3 Latin Square - Blocking in Two Orthogonal Directions

The data for this fabric manufacturer example are in a file called `Garments.jmp`. Each of four new fabric materials is being tested under each of four heat chamber positions, and due to time constraints, the manufacturer is limited to four runs. Fabric shrinkage is the response of interest.

Both `Run` and `Position` are the potential blocking sources of variation within the experiment design. The `Material` levels are assigned to runs and positions as a Latin square. A Latin square is a special design where the rows and columns have the same number of experimental units balanced within each other. Sudoku puzzles are Latin squares. The number of treatments is equal to the number of experimental units in the rows/columns and are randomized so that each treatment appears once in each row and column. Because runs and positions are a random selection of possible runs and positions, they should be considered random effects in our model. Materials are the only fixed effect the manufacturer is considering. These data are also analyzed in Littell et al. (2002).

A skeleton ANOVA for the Latin Square design follows. We have both the Run and Position blocking directions in the Experiment Design section. There is also the Run*Position intersection in the square, which is the experimental unit the Material treatments are randomized to.

Experiment Design		**Skeleton ANOVA**	
Source	***df***	**Source**	***df***
Run	4-1=3	Run	3
Position	4-1=3	Position	3
		Material	4-1=3
Run*Position	(4-1)*(4-1)=9	Residual	9-3=6
Total	16-1=15	Total	15

Mixed Model for a Latin Square

The statistical model for these data is

$$y_{ijk} = \mu + \alpha_i + r_j + p_k + e_{ijk}$$

α_i is the effect of the ith level of material.
r_j is the effect of the jth level of run and $r_j \sim N(0,\sigma_r^2)$.
p_k is the effect of the kth level of position and $p_k \sim N(0,\sigma_p^2)$.
e_{ijk} is the residual error and $e_{ijk} \sim N(0,\sigma^2)$.

JMP Instructions for Garment.jmp example

Click *Analyze* > *Fit Model* and select the *Standard Least Squares Personality*. Assign `Shrinkage` as the *Y* variable, and `Run`, `Position`, and `Material` as *Model Effects*. Designate `Run` and `Position` as random effects by selecting them and clicking *Attributes* > *Random Effect*. The completed model dialog box is shown below. Press *Run* to fit the model and generate a report.

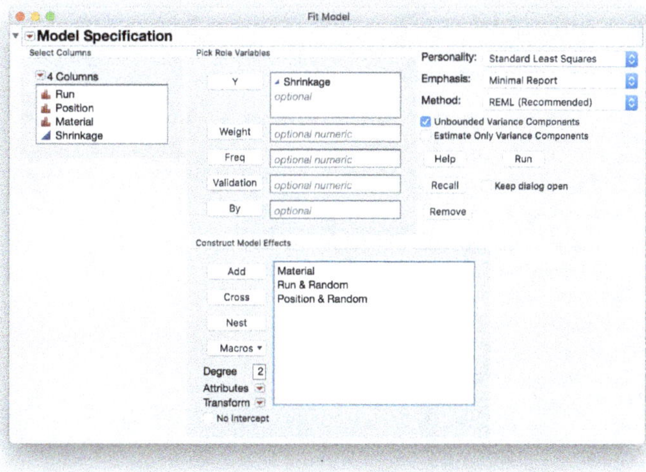

Results and Interpretation

Figure 4.1 shows the results of this analysis. The Actual by Predicted Plot shows predictions for the fixed effects only (no random effects included). The Residual by Predicted Plot does not show any problems with the residuals that might indicate non-constant variance.

The REML Variance Component Estimates table shows the estimates of σ_r^2, σ_p^2 and σ^2.

Figure 4.1: Standard Least Squares Report for the Garment Data

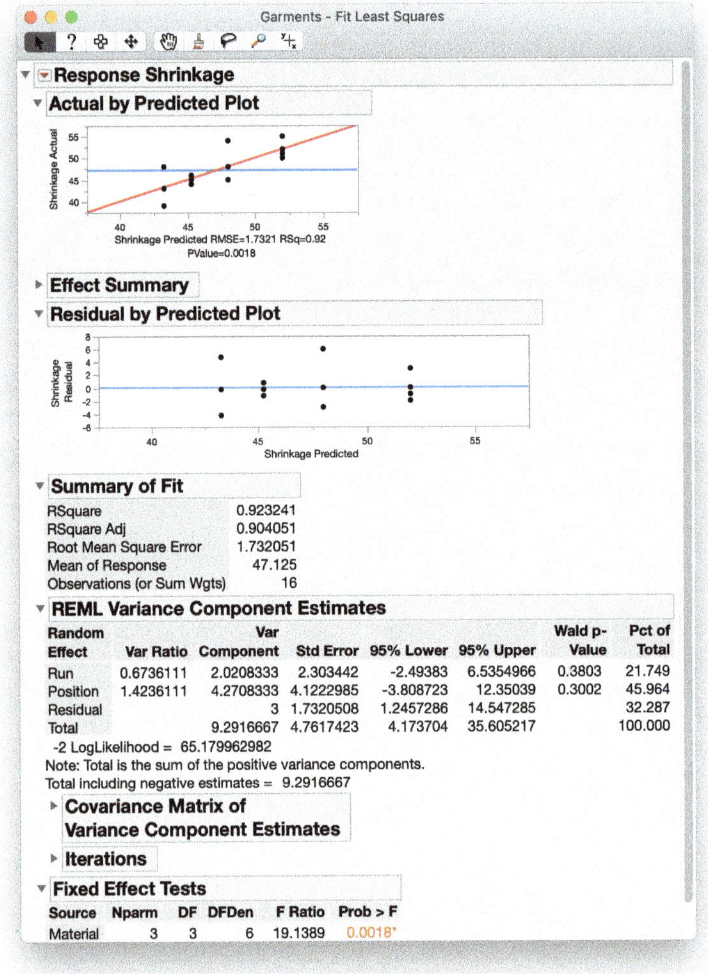

Summary of Fit

RSquare	0.923241
RSquare Adj	0.904051
Root Mean Square Error	1.732051
Mean of Response	47.125
Observations (or Sum Wgts)	16

REML Variance Component Estimates

Random Effect	Var Ratio	Var Component	Std Error	95% Lower	95% Upper	Wald p-Value	Pct of Total
Run	0.6736111	2.0208333	2.303442	-2.49383	6.5354966	0.3803	21.749
Position	1.4236111	4.2708333	4.1222985	-3.808723	12.35039	0.3002	45.964
Residual		3	1.7320508	1.2457286	14.547285		32.287
Total		9.2916667	4.7617423	4.173704	35.605217		100.000

-2 LogLikelihood = 65.179962982
Note: Total is the sum of the positive variance components.
Total including negative estimates = 9.2916667

Covariance Matrix of Variance Component Estimates

Iterations

Fixed Effect Tests

Source	Nparm	DF	DFDen	F Ratio	Prob > F
Material	3	3	6	19.1389	0.0018*

Blocking on **Run** and **Position** appears to have been effective.

The Fixed Effect Tests table shows that we have a significant effect of **Material** on **Shrinkage** with a p-value $p = 0.0018$. This only partially answers the researcher's question about the effect of the material on shrinkage. Shrinkage should be minimized, so the manufacturer wants to know the material that results in the least amount of shrinkage.

Explore the effect of material further by clicking *Estimates > Multiple Comparisons* option in the red triangle menu. A dialog box appears to choose the effect that you want to see as well as the multiple comparison method that you want to use. In this case, we select All Pairwise Comparisons with Tukey's HSD (honest significant difference). Figure 4.2 shows the completed Multiple Comparisons window for the interaction plot and the results for the pairwise LSMeans comparisons. Material B has the smallest amount of shrinkage with a Least Squares Means Estimate of 43.25. You can right-click in the Least Squares Means Estimates table and choose *Sort By Column > Estimate > Ascending* to display the least squares means in sorted order. The All Pairwise Differences table shows that material B has significantly less shrinkage than materials A and C, but not D.

Figure 4.2: Multiple Comparisons for the Garment Data

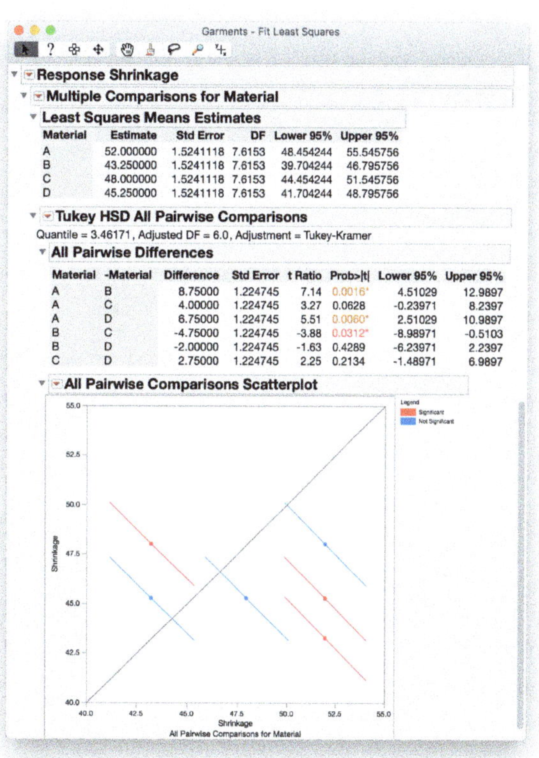

Material	Estimate	Std Error	DF	Lower 95%	Upper 95%
A	52.000000	1.5241118	7.6153	48.454244	55.545756
B	43.250000	1.5241118	7.6153	39.704244	46.795756
C	48.000000	1.5241118	7.6153	44.454244	51.545756
D	45.250000	1.5241118	7.6153	41.704244	48.795756

Material	-Material	Difference	Std Error	t Ratio	Prob>|t|	Lower 95%	Upper 95%
A	B	8.75000	1.224745	7.14	0.0016*	4.51029	12.9897
A	C	4.00000	1.224745	3.27	0.0628	-0.23971	8.2397
A	D	6.75000	1.224745	5.51	0.0060*	2.51029	10.9897
B	C	-4.75000	1.224745	-3.88	0.0312*	-8.98971	-0.5103
B	D	-2.00000	1.224745	-1.63	0.4289	-6.23971	2.2397
C	D	2.75000	1.224745	2.25	0.2134	-1.48971	6.9897

The All Pairwise Comparisons Scatterplot plots every pair of least squares means as a point, here a total of 4 choose 2 = 6 points. You can mouse over each point to interactively see which is which. The 95% confidence intervals of each difference are represented with lines perpendicular to the 45-degree reference line. Any interval that does not cross the 45-degree reference line has a Tukey HSD p-value less than 0.05. The further away a line is from the 45-degree reference, the more statistically significant the difference. You can verify this by comparing the plot with the numbers shown in the All Pairwise Differences table.

Materials B and D appear to be the best ones to use for minimal shrinkage.

JMP Pro Instructions for Garment.jmp Example

Using the Mixed Model personality of Fit Model, enter the one fixed effect, `Material`, on the Fixed Effects tab. Enter the two random effects, `Run` and `Position`, on the Random Effects tab. The completed model dialog box is shown below. Press *Run* to fit the model and generate a report.

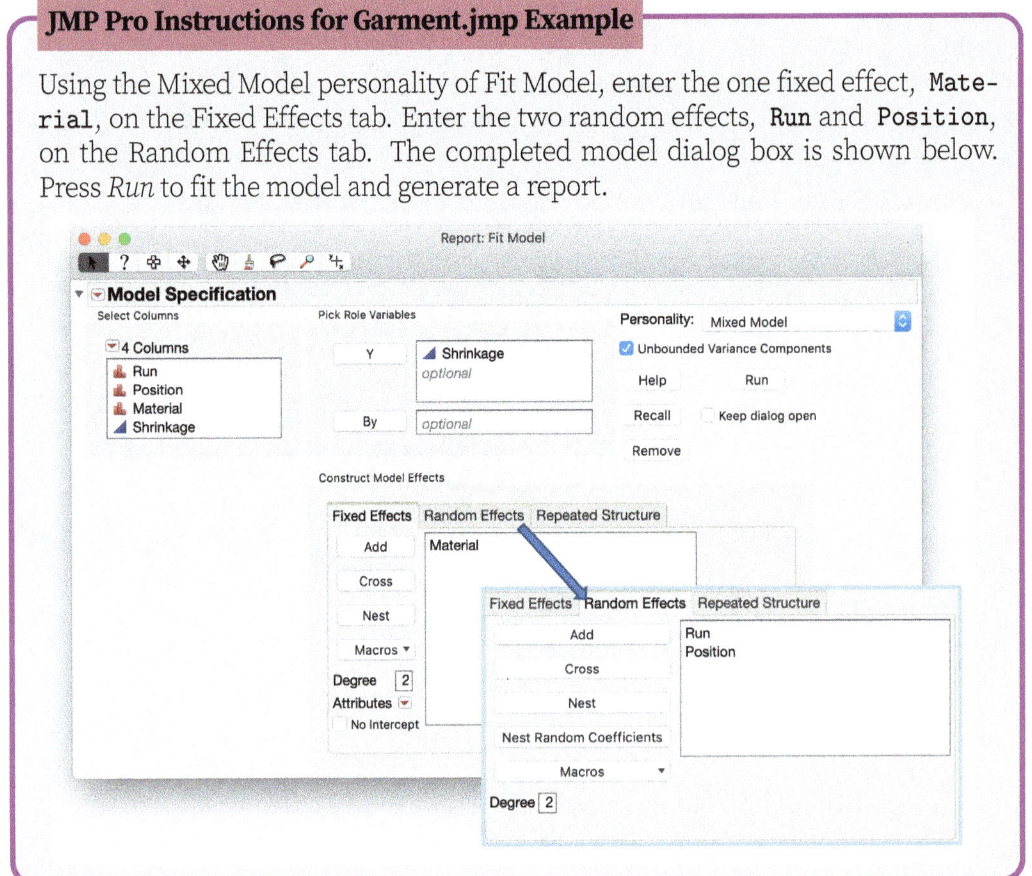

Results and Interpretation

Figure 4.3 shows the results of the analysis from the completed dialog box.

The Fit Mixed report shows both an Actual by Predicted Plot and an Actual by Conditional Predicted Report.

The Actual by Predicted plots the predicted values not accounting for the random effects associated with the actual values (same plot as previous analysis using regular JMP).

Figure 4.3: Fit Mixed Report for the Garment Data

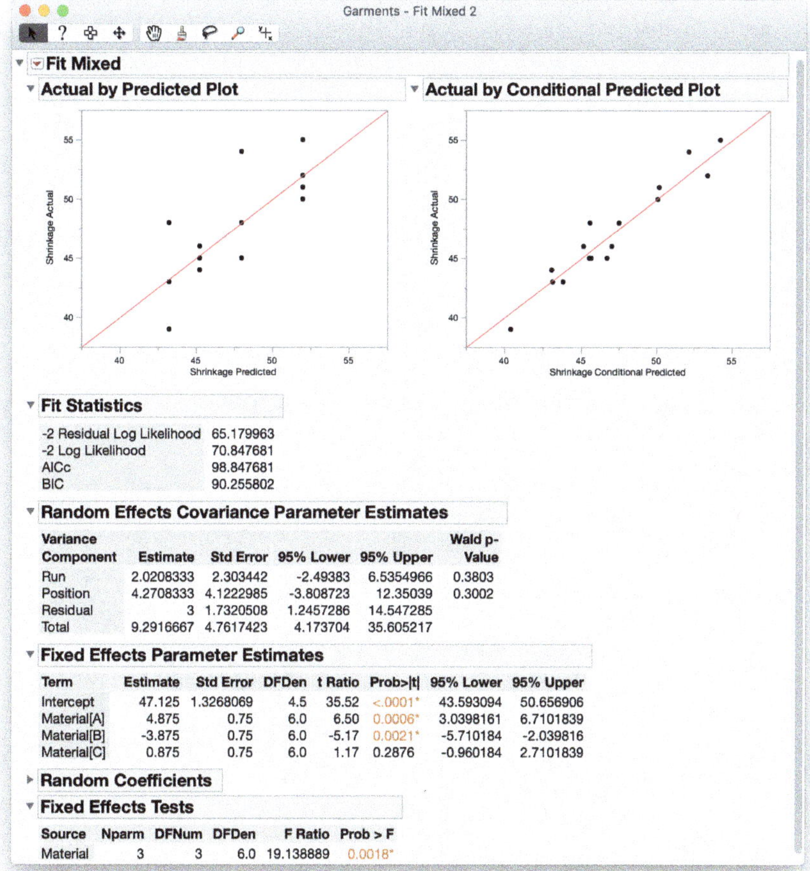

Fit Statistics

-2 Residual Log Likelihood	65.179963
-2 Log Likelihood	70.847681
AICc	98.847681
BIC	90.255802

Random Effects Covariance Parameter Estimates

Variance Component	Estimate	Std Error	95% Lower	95% Upper	Wald p-Value
Run	2.0208333	2.303442	-2.49383	6.5354966	0.3803
Position	4.2708333	4.1222985	-3.808723	12.35039	0.3002
Residual	3	1.7320508	1.2457286	14.547285	
Total	9.2916667	4.7617423	4.173704	35.605217	

Fixed Effects Parameter Estimates

Term	Estimate	Std Error	DFDen	t Ratio	Prob>|t|	95% Lower	95% Upper
Intercept	47.125	1.3268069	4.5	35.52	<.0001*	43.593094	50.656906
Material[A]	4.875	0.75	6.0	6.50	0.0006*	3.0398161	6.7101839
Material[B]	-3.875	0.75	6.0	-5.17	0.0021*	-5.710184	-2.039816
Material[C]	0.875	0.75	6.0	1.17	0.2876	-0.960184	2.7101839

Random Coefficients

Fixed Effects Tests

Source	Nparm	DFNum	DFDen	F Ratio	Prob > F
Material	3	3	6.0	19.138889	0.0018*

The Actual by Conditional Predicted plots the conditional predicted values. These are the predicted values *conditioned* upon the random effect, which means using the random effect solutions to calculate the predicted value. Note how the points in the conditional plot are somewhat closer to the 45-degree reference line, indicating visually that the random effects moderately improve the model fit.

Prediction Equations

Actual by Predicted plots use predictions from the formula $\hat{y}_i = \hat{\mu} + \hat{\alpha}_i$, which are known as the marginal predicted values. The conditional predicted values use the formula $\hat{y}_{ijk} = \hat{\mu} + \hat{\alpha}_i + \hat{r}_j + \hat{p}_k$; these predictions include the random effect solutions.

The Fixed Effects Parameter Estimates table shows the estimates of the fixed effects, which are used to calculate the Least Squares Means. The coefficients in this table correspond to a sum-to-zero parameterization of the nominal effects. Coefficients corresponding to an indicator parameterization of the nominal effect are available from the *Model Reports > Indicator Parameterization Estimates* option in the red triangle menu. The indicator parameterization utilizes dummy 0-1 variables for all levels except the last one, and it is the parameterization commonly used in SAS linear and mixed modeling procedures with effects specified in a CLASS statement. In JMP there is no such assignment in the model dialog box, rather we pre-assign all variables to be Continuous or Nominal in the JMP data table, and these attributes are automatically incorporated in the model.

You can obtain Multiple Comparisons from the report's red triangle menu. It is the same report as in the Standard Least Squares personality, so it is not shown here again. The multiple comparisons are shown in Figure 4.2, and all conclusions are the same.

4.4 Mouse Condition: Negative Block Variance Example

Figure 4.4: Experiment Design for Mouse Condition

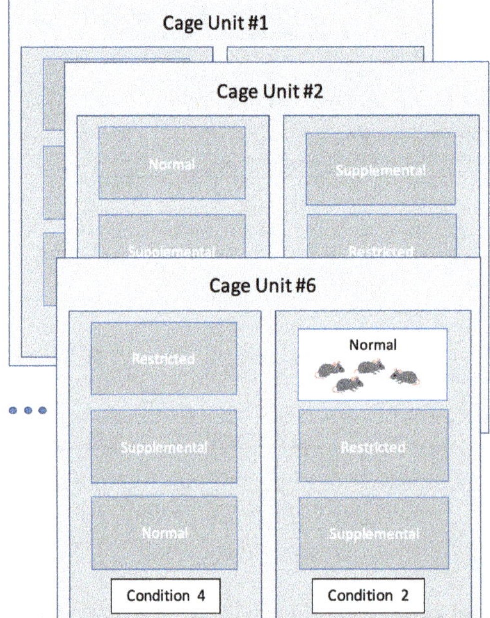

A trial involving laboratory mice investigates the effects of four housing conditions (CONDITION 1, 2, 3, and 4) and three feeding regimens (DIET "restricted," "normal," and "supplemental") on weight gain in the mice. As described in Stroup et al. (2018), two housing conditions can be accommodated in one CAGE unit, and within each housing-condition-in-a-caging-unit, mice can be separated into the three distinct diet groups.

CAGE is a blocking factor for incomplete blocks. CONDITION is the whole-plot treatment factor being applied within those CAGE halves. Those CAGE halves can be identified by specifying the CAGE*CONDITION, so this identifies the whole plot experimental unit. DIET is the factor applied at the split-plot experimental unit, which is identified by specifying the CAGE*CONDITION*DIET.

In the Skeleton ANOVA, Half-Cage identifies the whole-plots to which Conditions are randomized. Once the Condition has been assigned, these can be uniquely identified by the Cage-Condition combination. Cage-Sixth refers to the split-plot thirds that are within the Half-Cages (thus 1/6 of a cage). There are 3-1=2 *df* per Half-Cage and 12 Half-Cages total.

Experiment Design		**Skeleton ANOVA**	
Source	***df***	**Source**	***df***
Cage	6-1=5	Cage	5
		Condition	4-1=3
Half-Cage	(2-1)*6=6	Cage*Condition	6-3=3
		Diet	3-1=2
		Condition*Diet	3*2=6
Cage-Sixth	(3-1)*12=24	Residual	24-2-6=16
Total	36-1=35	Total	35

Mixed Model for Split-Plot within Blocks

The statistical model for these data is

$$y_{ijk} = \mu + c_i + \alpha_j + (c\alpha)_{ij} + \beta_k + \alpha\beta_{jk} + e_{ijk}$$

y_{ijk} is the weight gain of the ijk^{th} unit of mice.
c_i is the effect of the jth cage unit and $r_i \sim N(0,\sigma_r^2)$.
α_j is the effect of the ith condition.
$(c\alpha)_{ij}$ is the whole-plot experimental unit, the half-cage-unit, which is equivalently the effect of the ith level of cage and the jth level of condition, where $(c\alpha)_{ij} \sim N(0,\sigma^2_{(c\alpha)})$.
β_k is the effect of the kth diet.
$\alpha\beta_{jk}$ is the interaction between the jth condition and the kth diet.
e_{ijk} is the residual error and $e_{ijk} \sim N(0,\sigma^2)$.

JMP Instructions for MouseCondition.jmp Example

Click *Analyze > Fit Model* and select the *Standard Least Squares Personality*. Assign `Gain` as the *Y* variable. Select `Condition` and `Diet` and click the *Macros > Full Factorial* to include the main effects and the interaction effect in the *Model Effects* box. Also add `Cage` to the *Model Effects* box and select `Cage` and `Condition` in the *Columns* list and click *Cross* to add their interaction to the *Model Effects* box.

Designate `Cage` and `Cage*Condition` as random effects by selecting them and clicking *Attributes > Random Effect*. The completed model dialog box is shown below. Press *Run* to fit the model and generate a report.

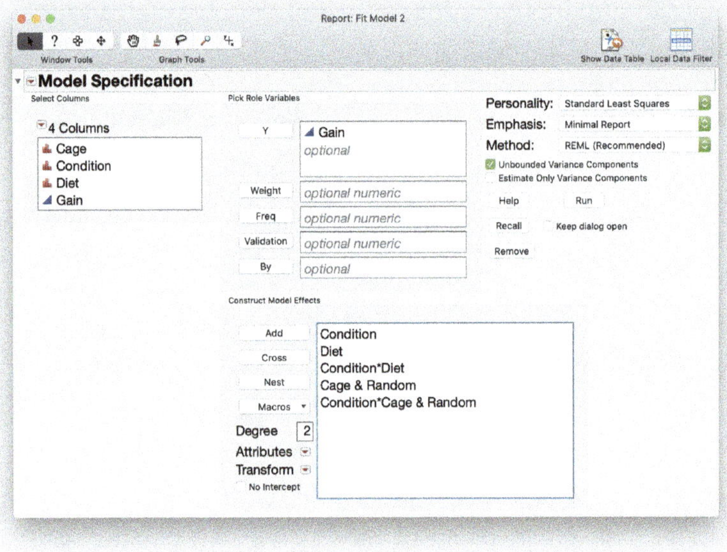

Results and Interpretation

Figure 4.5 shows the results of the analysis. The Fit Least Squares report shows, in the Fixed Effect Tests output, that the Condition effect has the most evidence for a statistically significant difference, with a p-value of 0.07982, but that none of the fixed effects show statistically significant differences in this model at the $\alpha = 0.05$ level.

However, we also see, in the REML Variance Component Estimates, that we have a *negative* estimate for the variance of Condition*Cage. What does this mean, and how does this impact our fixed effect comparisons?

Negative Variance Component Estimates

This is not a mistake or, necessarily, a problem.

Figure 4.5: Fit Least Squares Report for the Mouse Condition Data

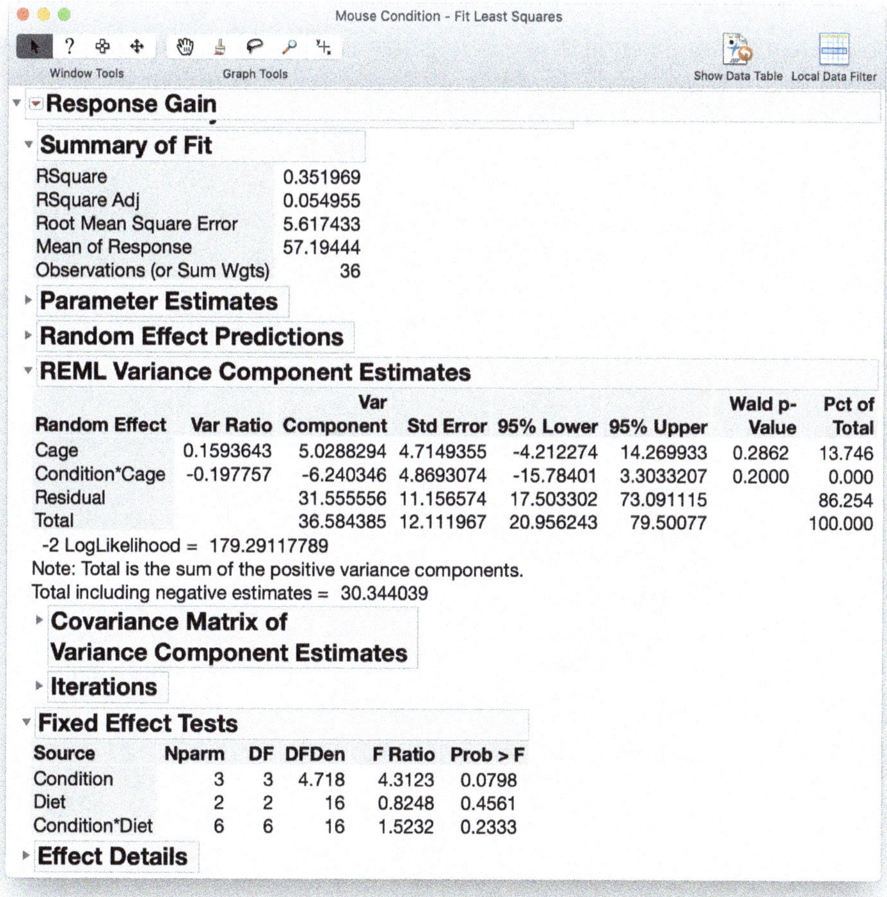

Response Gain

Summary of Fit

RSquare	0.351969
RSquare Adj	0.054955
Root Mean Square Error	5.617433
Mean of Response	57.19444
Observations (or Sum Wgts)	36

Parameter Estimates

Random Effect Predictions

REML Variance Component Estimates

Random Effect	Var Ratio	Var Component	Std Error	95% Lower	95% Upper	Wald p-Value	Pct of Total
Cage	0.1593643	5.0288294	4.7149355	-4.212274	14.269933	0.2862	13.746
Condition*Cage	-0.197757	-6.240346	4.8693074	-15.78401	3.3033207	0.2000	0.000
Residual		31.555556	11.156574	17.503302	73.091115		86.254
Total		36.584385	12.111967	20.956243	79.50077		100.000

-2 LogLikelihood = 179.29117789

Note: Total is the sum of the positive variance components.

Total including negative estimates = 30.344039

Covariance Matrix of Variance Component Estimates

Iterations

Fixed Effect Tests

Source	Nparm	DF	DFDen	F Ratio	Prob > F
Condition	3	3	4.718	4.3123	0.0798
Diet	2	2	16	0.8248	0.4561
Condition*Diet	6	6	16	1.5232	0.2333

Effect Details

Negative variance component estimates might happen when, for example, a variance is very small or when there is negative correlation among experimental units. This latter situation often happens when there is competition for resources among plots or units, as here with mice in a cage. In such cases, the REML optimal value for the variance estimate can cross into the negative region.

It might be tempting to go back to the dialog box and deselect the *Unbounded Variance Components* option on the right panel, to force all variance components to be nonnegative. We suggest that you resist this temptation. You can read more about this issue in

Section 5.7 of Stroup et al. (2018), but, in summary, you get better control over Type 1 error for your fixed effects comparisons, and in some cases better power, by allowing negative variance component estimates.

Although it might seem strange to report negative variance component estimates, this unbounded fit is the best model for estimating the fixed effects comparisons. Indeed, theoretically variances can never be negative, but the mixed model is actually modeling covariances within a block, and these can be negative, and correspond to negative intraclass correlations. We suggest, after you confirm that your model is defined as you intended and that the negative estimate at least makes some practical sense in the conditions that you are studying, you include caveat language around the negative variance component estimates, such as we provide in the Methods and Results Summary for Publication section at the end of this chapter, and proceed with your analysis using the unbounded model.

Figure 4.6: Fit Least Squares Means for the Mouse Condition Data

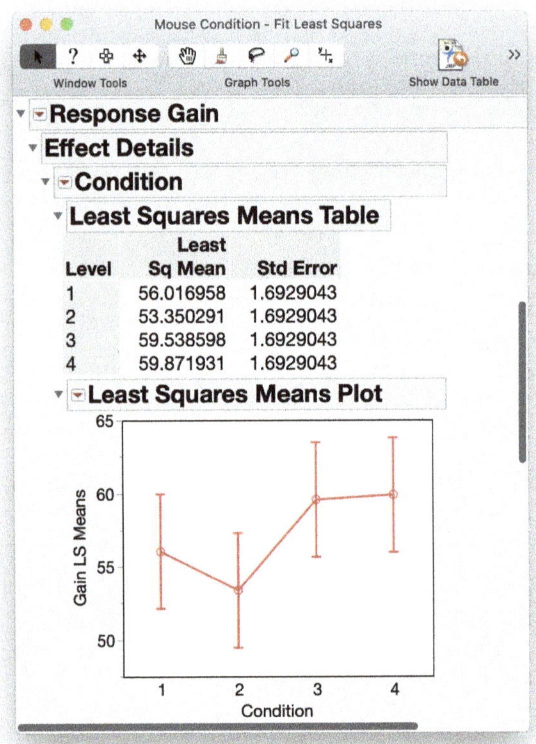

Comparisons and Contrasts

Although we have a borderline p-value for the Condition effect, we might still have planned comparisons to investigate within that effect. For interest in the Tukey HSD comparisons for all pairs, we do not find anything to be significant. However, we might want to make a planned non-pairwise comparison. For example, we might have decided at the outset of the experiment to compare Condition 2 to the average of the other three conditions. We can do this by selecting the red triangle next to the Condition heading in the Effect Details, and choose *LSMeans Contrast*. Click on the plus button to choose the groups for one side of the contrast (Condition2, in this case), and click on the minus button to choose the groups for the other group (Conditions 1, 3, and 4). From Figure 4.7 we can see that the mean Gain for the Condition 2 group is statistically significantly different from the average of the other three groups. The average gain from the other three groups combined is higher than the average gain from Condition 2.

Figure 4.7: Contrasts for the Mouse Condition Data

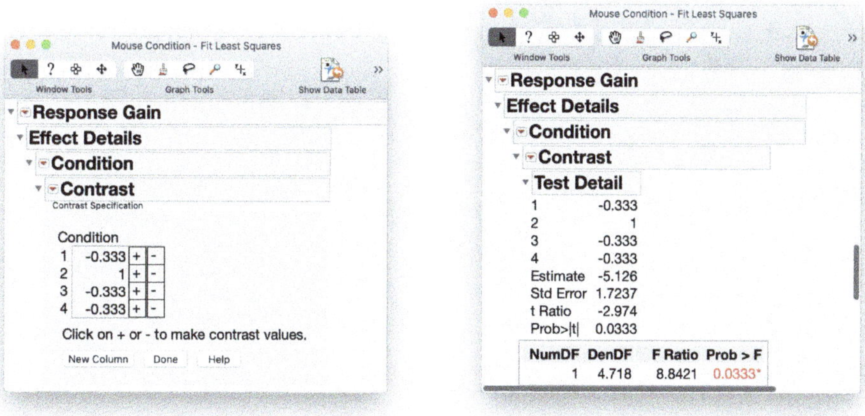

These comparison and contrast options should not be used merely to search for significance after data collection. When relying on significance values in hypothesis testing as evidence of important effects, these should be used carefully and based on planned comparisons in general. See Leung (2011) for a discussion of the use of *post hoc* tests.

Generating SAS Code from JMP

Corresponding SAS code can be generated from the Standard Least Squares personality in Fit Model. While the dialog box is open, after filling out the *Y* variable and the *Model Effects*, click on the red triangle at the top left by the **Model Specification** heading. Select *Create SAS Job* to generate a new script window with the SAS code corresponding to the JMP model just specified in the dialog box.

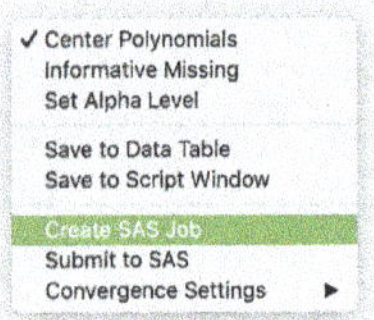

The following code is generated. (The DATA step has been abbreviated to save space.) You can now adapt this code and copy it to run in SAS or run it directly in SAS through this window in JMP if you have set up your JMP to SAS connection. See the online JMP Help at https://www.jmp.com/help and search "SAS" for more information about running SAS through JMP.

```
DATA Mouse_Condition; INPUT  Cage Condition Diet &$ Gain; Lines;
1  1  normal  58
1  1  restrict  58
1  1  suppleme  58
1  2  normal  54
1  2  restrict  46
1  2  suppleme  57
2  1  normal  49
2  1  restrict  50
2  1  suppleme  57
6  3  normal  65
6  3  restrict  64
6  3  suppleme  54
6  4  normal  59
6  4  restrict  47
6  4  suppleme  73
;
RUN;

PROC MIXED ASYCOV NOBOUND  DATA=Mouse_Condition ALPHA=0.05;
CLASS Cage Condition Diet;
MODEL  Gain =  Condition Diet Condition*Diet/ SOLUTION DDFM=KENWARDROGER
RANDOM Cage Condition*Cage / SOLUTION ;
RUN;
```

JMP Pro Instructions for Mouse Condition.jmp Example

Select **Gain** as the *Y* variable and select the Mixed Model personality of Fit Model.

On the Fixed Effects tab, enter the two fixed effects and their interaction by selecting **Condition** and **Diet** and selecting the *Macros* button and then *Full Factorial*.

Enter the two random effects, **Cage** and **Cage*Condition**, on the Random Effects tab. (To enter the interaction, select **Cage** in the Random Effects box, select **Condition** in the columns list, and click the *Cross* button.)

The completed model dialog box is shown below. Press *Run* to fit the model and generate a report.

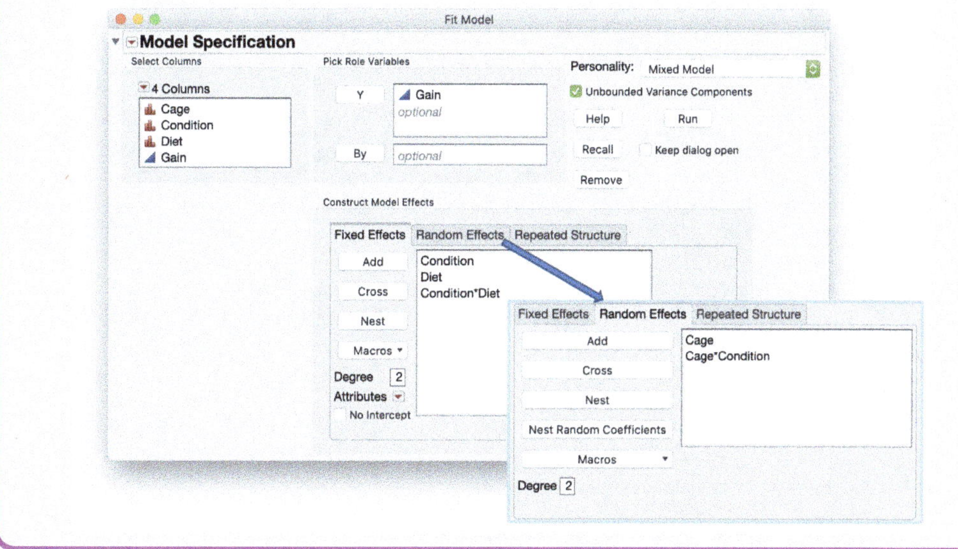

Results and Interpretation

Figure 4.8 shows the results of the analysis from the completed dialog box.

We see the same negative variance component estimate for the whole plots, Cage*Condition. Again we allow this in order to retain our expected Type 1 error rate and potentially improve power over the model with only positive-bounded estimates.

Figure 4.8: Fit Mixed Report for the Mouse Condition Data

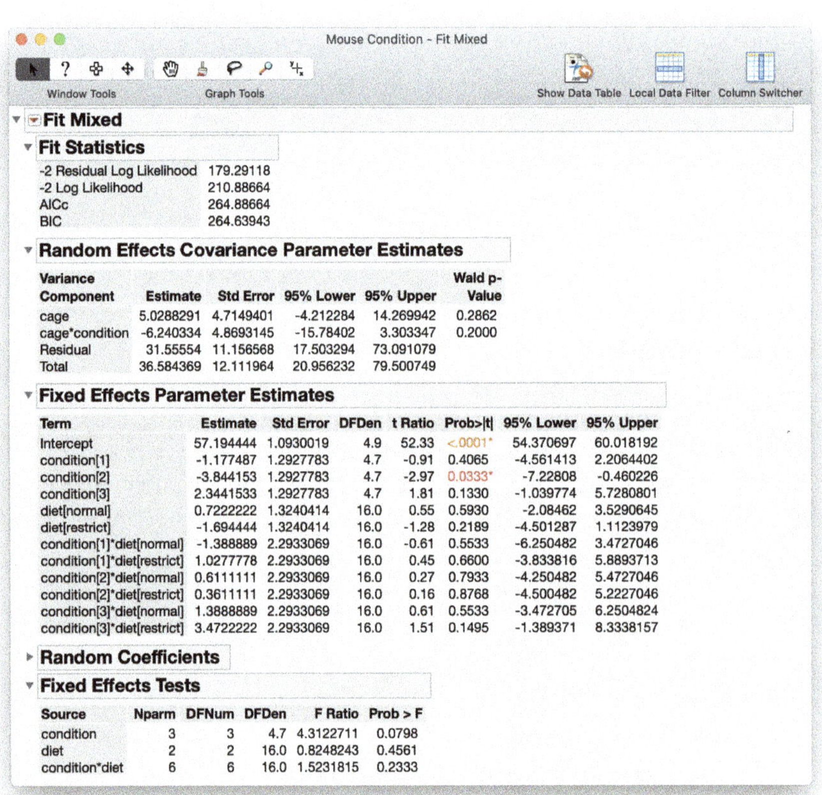

Mouse Condition - Fit Mixed

Fit Mixed

Fit Statistics

-2 Residual Log Likelihood	179.29118
-2 Log Likelihood	210.88664
AICc	264.88664
BIC	264.63943

Random Effects Covariance Parameter Estimates

Variance Component	Estimate	Std Error	95% Lower	95% Upper	Wald p-Value
cage	5.0288291	4.7149401	-4.212284	14.269942	0.2862
cage*condition	-6.240334	4.8693145	-15.78402	3.303347	0.2000
Residual	31.55554	11.156568	17.503294	73.091079	
Total	36.584369	12.111964	20.956232	79.500749	

Fixed Effects Parameter Estimates

Term	Estimate	Std Error	DFDen	t Ratio	Prob>|t|	95% Lower	95% Upper
Intercept	57.194444	1.0930019	4.9	52.33	<.0001*	54.370697	60.018192
condition[1]	-1.177487	1.2927783	4.7	-0.91	0.4065	-4.561413	2.2064402
condition[2]	-3.844153	1.2927783	4.7	-2.97	0.0333*	-7.22808	-0.460226
condition[3]	2.3441533	1.2927783	4.7	1.81	0.1330	-1.039774	5.7280801
diet[normal]	0.7222222	1.3240414	16.0	0.55	0.5930	-2.08462	3.5290645
diet[restrict]	-1.694444	1.3240414	16.0	-1.28	0.2189	-4.501287	1.1123979
condition[1]*diet[normal]	-1.388889	2.2933069	16.0	-0.61	0.5533	-6.250482	3.4727046
condition[1]*diet[restrict]	1.0277778	2.2933069	16.0	0.45	0.6600	-3.833816	5.8893713
condition[2]*diet[normal]	0.6111111	2.2933069	16.0	0.27	0.7933	-4.250482	5.4727046
condition[2]*diet[restrict]	0.3611111	2.2933069	16.0	0.16	0.8768	-4.500482	5.2227046
condition[3]*diet[normal]	1.3888889	2.2933069	16.0	0.61	0.5533	-3.472705	6.2504824
condition[3]*diet[restrict]	3.4722222	2.2933069	16.0	1.51	0.1495	-1.389371	8.3338157

Random Coefficients

Fixed Effects Tests

Source	Nparm	DFNum	DFDen	F Ratio	Prob > F
condition	3	3	4.7	4.3122711	0.0798
diet	2	2	16.0	0.8248243	0.4561
condition*diet	6	6	16.0	1.5231815	0.2333

The Fit Mixed output allows us to explore the marginal and conditional model predictions. The marginal model inference averages the effects over all levels of the random effects. The conditional model inference allows us to specify a level for the random effects and see the predictions conditional on that selection. Choose the *Marginal Model Inference* from the red triangle next to the Fit Mixed heading, and select the *Profiler*. Do the same for the *Conditional Model Inference > Profiler*. These are shown in Figure 4.9. In the marginal profiler, we see that the average Gain for the Restricted Diet in Condition 3 is 61.32, with a 95% Confidence Interval on that mean of (54.82,67.81). In the conditional profiler, we see that this prediction can be further conditioned on the cage. For the Restricted Diet with Condition 3 in Cage 5, that expected mean Gain is 59.13.

The Marginal Model Inference more closely matches the intention of modeling cage as random. For a random effect for cage, we typically are not interested in making predictions within a specific cage. However, the Conditional Model Inference allows us to examine if there are dramatic differences in the predictions within one cage compared to another, providing a more visual method to identify unusual cage effects, besides just reading the BLUPs from the Random Coefficients table.

Figure 4.9: Fit Mixed Profilers for the Mouse Condition Data

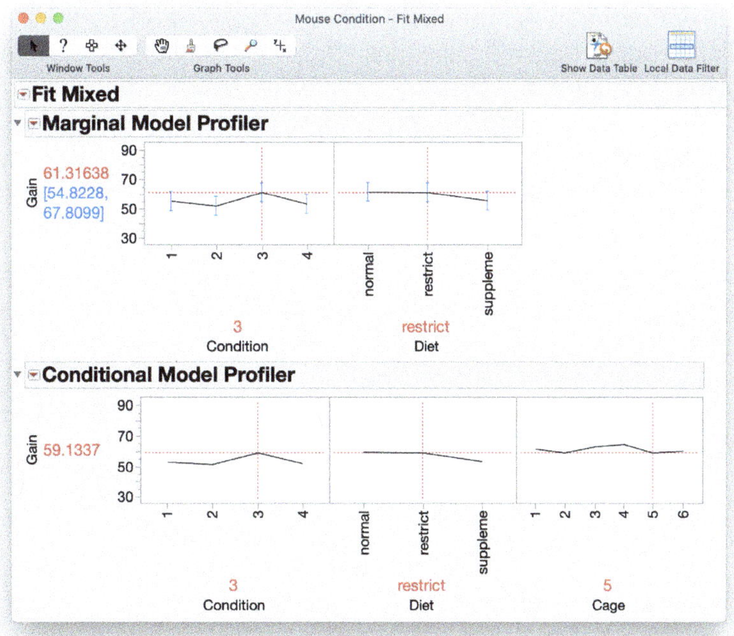

Methods and Results Summary for Publication

Methods

A split-plot mixed model is fit in JMP Pro, version 16. The model is specified with Condition, Diet, and the Condition-by-Diet interaction as fixed effects and with Cage and Cage*Condition as random effects on the outcome Gain. Unbounded variance components are estimated using REML.

Results

The Cage variance estimate of 5.03 is less than a sixth of the magnitude of the residual error, indicating that only a small amount of additional variance is explained by the cage-to-cage differences. The Condition*Cage variance (the Whole Plot error) estimate is -6.24. This negative variance component estimate likely indicates that the whole plot error is negligible (and therefore estimated below zero to improve overall model fit) or a result of negative correlation among observations sharing the same whole plot, which might exist due to resource competition or other forces. Allowing the variance components to be estimated unbounded improves our expected Type 1 error rate, so we do not refit the model to force nonnegative variance component estimates.

None of the fixed effects are statistically significant at the $\alpha = 0.05$ level, however, the contrast of Condition 2 compared to the average of the other three conditions showed that Condition 2 results in statistically significantly lower Gain. We conclude that Condition 2 is not as effective on Gain as the other three conditions, when compared across all Diets together. The contrast of the Restricted Diet in Condition 3 versus the average of the other three conditions for that diet shows a statistically significant increase in Gain for the Restricted Diet in Condition 3.

From these two planned contrasts, we conclude that Condition 2 should be excluded in future studies due to its subpar Gains, and Restricted Diet seems to result in the highest Gains for Condition 3. Further investigation on Supplemental Diet in Condition 4 is also recommended; however, we do not test specifically for this because it was not a planned contrast.

4.5 Exercises

1. For the Mouse Condition in Section 4.4, create a contrast to compare, within the Restricted Diet, the gains in Condition 3 versus the average of the other Conditions by selecting the red triangle next to the Condition heading in the Effect Details of the Fit Least Squares output, and choose *LSMeans Contrast*.
2. In an extension of the semiconductor experiment introduced in Chapter 3, the engineer again has several process conditions that could affect resistance on the wafers produced. The position of the chips on the wafer can also affect resistance. There is

also potential variability in the process due to the position (another blocking factor). The engineer wants to minimize resistance and understand whether position of the chip on the wafer affects resistance.

(a) Write out the new statistical model that includes this new blocking factor.
(b) Add this second blocking factor to the model and compare the results on the new model to the results without this additional blocking factor.

Chapter 5

Regression, Random Coefficients, and Multilevel Models

This chapter introduces the first models with more complex covariance structures. That is, the various error terms are not necessarily independent from one another but have some covariance between them. A consequence of this complexity is that these models can only be fit using the Mixed Model personality in JMP Pro. However, we will hope to show that a "complex" model need not be difficult or confusing.

5.1 Motivating Examples

Stability Trial — Drug developers perform a stability trial to determine optimal shelf life of a pharmaceutical. The batches used to measure the strength over time are randomly selected from possible batches created.

Student Achievement — Eighth grade students take mathematics achievement tests. Researchers are interested in not only any overall trend among all students but also trends within individual classrooms.

5.2 Conceptual Background

Most readers of this book are familiar with simple linear regression. We have some predictor variable observed on at a variety of values and we measure an outcome. We want to describe the relationship between the predictor and the outcome using a line, $y=mx+b$. The slope, m, and intercept, b, are estimated from the x and y data observed. We are not limited to a single predictor variable, but we will keep it simple as we expand on this idea to a random coefficient model.

In the stability trial the full data could be used to fit the simple linear regression, however, the batches of the drug may have different slopes or intercepts if they were fit individually. Which line is "correct"? The overall line, a particular individual batch's line, or something else entirely? A random coefficient model allows all of the data to be used *and* individual batch lines to be estimated at the same time. The approach produces intermediate solutions, shrinking the individual line fits to the global one in an optimal way, and borrows strength across the whole data. This maximizes the information available when estimating each line. The key assumption in the random coefficient model is that the individual intercepts and slopes are a random sample from the population of possible coefficients. The terminology and models for the random coefficient model comes from the statistics literature (Laird and Ware (1982), Rutter and Elashoff (1994) and Wolfinger (1996)).

Key Terminology

Random Coefficient A regression coefficient for one or more covariate that is assumed to be a random sample from a population of possible coefficients.

Random Coefficient Model A model where one or more regression coefficients are random coefficients. Usually, the random coefficients are assumed to be correlated.

At the same time, the social sciences have long had data with a nested or hierarchical structure. In our example above, students are nested within schools (which could be nested in districts, which could be nested in states). *Hierarchical* or *multilevel linear models* have been studied extensively over the years, with foundational books by Raudenbush and Bryk (2002) and Goldstein (1987). In a popular paper from 1998, Singer shows that the HLM from the social sciences is a mixed model, specifically a random coefficient model, and could be fit using SAS PROC MIXED (Singer, 1998). In the tradition of Singer, we show how to fit this traditional HLM as a random coefficient model in JMP Pro.

Many times when using a random coefficient model, researchers are interested not only in the fixed effects estimates but also in the random subject effects. Predictions using the observed random effects are referred to as conditional predictions; the predictions are *conditioned* on a level of the random effect. Predictions using only the fixed effects are marginal predictions. We mentioned briefly in Chapter 4 the differences in the marginal and conditional prediction equations. The conditional predictions also are referred to as BLUPs, Best Linear Unbiased Predictors.

The terminology BLUP comes from Henderson (1963) in his seminal work developing procedures for predicting breeding values of randomly selected sires in animal genetics experiments. The theoretical properties of Henderson's BLUPs were shown by Harville (1976). BLUPs can also be categorized as a form of empirical Bayes estimator under a squared (L2) norm (Robinson, 1991).

5.3 Stability Trial

The **`Stability Trial.jmp`** data follow the International Conference on Harmonization's Q1E (2003) guidelines for such trials. The guidelines state that a minimum of three batches of the drug must be observed over time. Suggested times of observation are 0, 3, 6, 9, 12, and 18 months. Additional observations are sometimes taken at 24 and 36 months. These data have observations through 24 months. The response variable is normalized so that the mean stability at time zero is 100. If the stability limiting characteristic is decreasing, 90 is usually set as the lower limit; if it is increasing, 110 is typically the upper limit. Due to the initial normalization of the response, these limits represent a 10% change from "stable." The shelf life is then determined by the time at which the characteristic crosses the upper or lower limit.

The Q1E guidelines specify the statistical model as follows.

Q1E Stability Trial Model

$$y_{ij} = \beta_0 + \delta_{0i} + (\beta_1 + \delta_{1i})t_{ij} + e_{ij}$$

y_{ij} is the continuous response variable.
t_{ij} is the time of observation.
β_0 and β_1 are the overall intercept and slope.
δ_{0i} and δ_{1i} are the batch-specific intercept and slope of the ith batch.
e_{ij} is the residual error and $e_{ij} \sim N(0, \sigma^2)$.

Notice that this model is similar to the "usual" line but adds the additional δ terms for the batches. This is an Analysis of Covariance (ANCOVA) model to test the equality of batches. The guidelines specify that the batch effects are fixed effects with procedures to test for equality among batches then estimating the shelf life. However, if we remember our definition of fixed effects, the conclusions that we draw from this model apply to "these" batches and not to future batches, which is the obvious goal of the study. If the batch effects are treated as random, then inference applies to the future batches as well. The model is modified like this.

Random Coefficient Model for the Stability Trial

$$y_{ij} = \beta_0 + b_{0i} + (\beta_1 + b_{1i})t_{ij} + e_{ij}$$

y_{ij} is the continuous response variable.
t_{ij} is the time of observation.
β_0 and β_1 are the overall intercept and slope.
b_{0i} and b_{1i} are the batch-specific intercept and slope of the ith batch where

$$\begin{bmatrix} b_{0i} \\ b_{1i} \end{bmatrix} \sim N\left(\begin{bmatrix} 0 \\ 0 \end{bmatrix}, \begin{bmatrix} \sigma_0^2 & \sigma_{01} \\ \sigma_{01} & \sigma_1^2 \end{bmatrix}\right)$$

e_{ij} is the residual error and $e_{ij} \sim N(0, \sigma^2)$.

Besides the variances associated with the random intercepts and slopes, the model includes a covariance, σ_{01}, between the intercepts and slopes. This covariance is included because quite often batches with higher intercepts also have higher slopes (and vice versa), which implies they are correlated with each other. The covariance also makes the model fully location and scale invariant with respect to transformations of the response and time. We can visualize linear fits using Graph Builder with the Stability Trial data.

Graph Builder Instructions for Stability Trial

With the **`Stability Trial.jmp`** data table open, go to *Graph* > *Graph Builder*. Drag **`y`** to the *Y* drop zone. Drag **`month`** to the *X* drop zone. Graph Builder defaults to fitting a smoothing spline to the data. We want a linear fit, so click the line with points button for Line of Fit. The confidence band is not necessary for this visualization, so uncheck the box next to *Fit* under the Line of Fit options.
Finally, we want to see the difference in batches, so drag **`batch`** to the *Overlay* drop zone. This colors the points by batch and fits individual lines for each batch. Click *Done* to complete the graph.

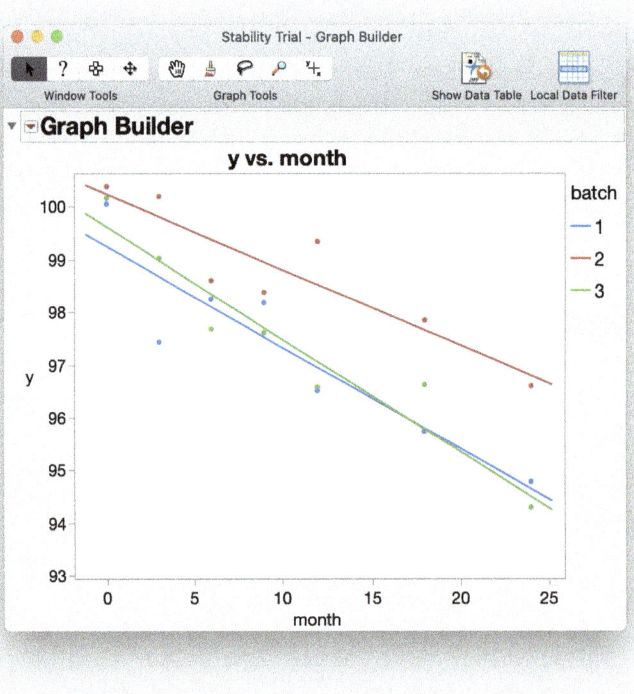

From the graph we can see that batch number two has a higher intercept and appears to have a slower decrease than batches one or three. We can formally test whether this apparent correlation between the intercept and slope is statistically supported by the data when we fit the random coefficient model.

Note: Although this example uses a drug study to demonstrate the random coefficient model, if your study falls under regulatory guidelines such as the Q1E, then you must follow those guidelines. There are other studies with similarities to drug trials without the same regulatory requirements that could use the random coefficient model for better results.

Random Coefficient Instructions for Stability Trial

With the **`Stability Trial.jmp`** data table open, go to *Analyze > Fit Model*. Enter **`y`** as the *Y* variable. Change the default personality from *Standard Least Squares* to *Mixed Model*.
Select **`month`** in the *Select Columns* box then click *Add* to add fixed effect of month to the *Construct Model Effects* box.
Enter **`month`** in the *Random Effects* tab of the *Construct Model Effects* box. Then select **`batch`** in the *Select Columns* box and **`month`** in the *Random Effects* tab and click *Nest Random Coefficients* to create the batch-specific random intercept and slope.
Click *Run* to fit the model.

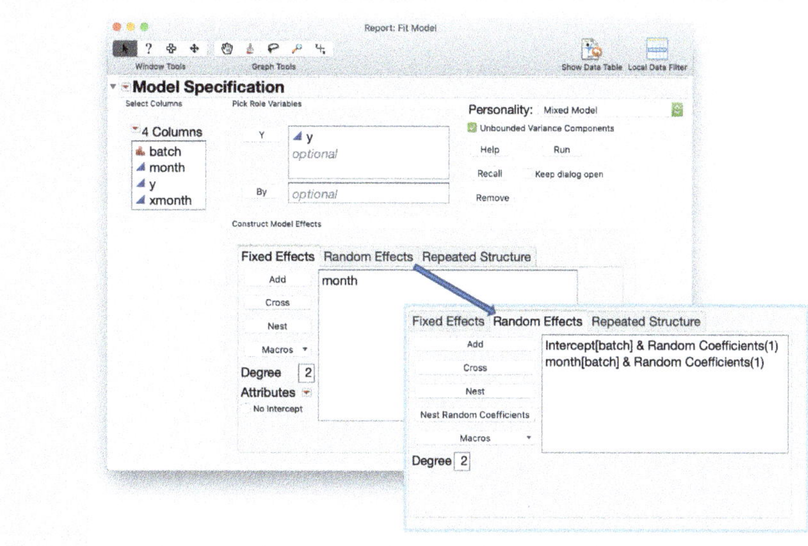

Results and Interpretation

In Figure 5.1, we see a good visual fit in the Actual by Predicted Plots, particularly the batch-specific predictions fit well as shown in the Actual by Conditional Predicted Plot.

The Random Effects Covariance Parameter Estimates table shows the intercept and slope variances and covariance with Wald confidence intervals. The confidence intervals indicate the variances and covariances are not different from zero. Wald intervals for variance components, especially with few batches, can tend to be too wide, so after removing the covariance from the model, we can compare likelihoods to verify if the simplified model is better. To maintain the goal of "all future batches", we need to keep the random intercept and slope, but we can simplify the model to remove the covariance.

Figure 5.1: Fit Mixed Report for the Stability Trial Data

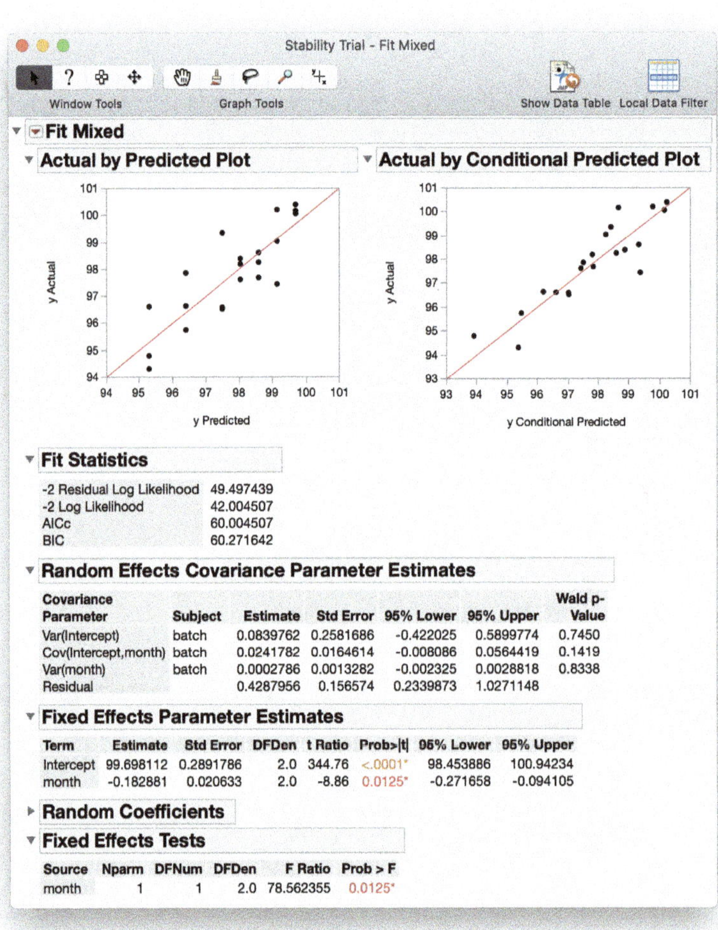

Simplified Model Instructions for Stability Trial

With the `Stability Trial.jmp` data table open, go to *Analyze > Fit Model*. Enter `y` as the *Y* variable. Change the default personality from *Standard Least Squares* to *Mixed Model*.

Select `month` in the *Select Columns* box then click *Add* to add fixed effect of month to the *Construct Model Effects* box.

Enter `batch` and `month` in the *Random Effects* tab of the *Construct Model Effects* box. Then select `batch` in the *Select Columns* box and `month` in the *Random Effects* tab and click *Nest* to create the batch-specific random intercept and slope. Choosing Nest rather than Nest Random Coefficients eliminates the covariance term.

Click *Run* to fit the model.

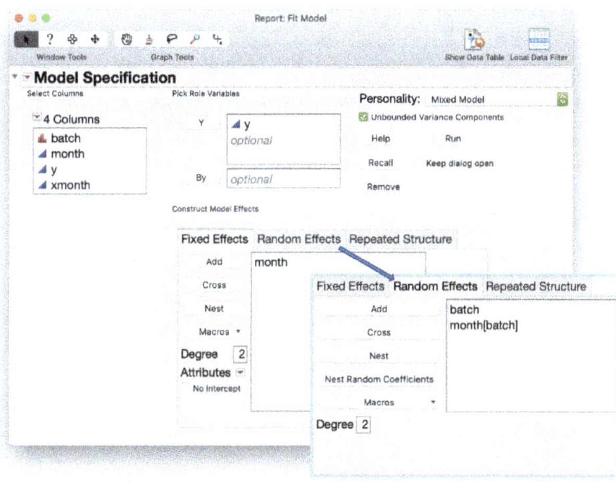

Results and Interpretation

The results of the simplified model are shown in Figure 5.2. The -2 Res Log Likelihood of the full random coefficient model was 49.497 and the -2 Res Log Likelihood for the reduced model is 52.5467. This leads to a likelihood ratio χ^2 statistic of 3.05, which is less than the critical value for a χ^2 with 1 *df*, 3.84. This indicates the covariance term is not necessary for the model.

The Fixed Effects Parameter Estimates give the overall intercept and slope estimates. These are known as the *population-average* estimates and are the same as if we had not included the random coefficients in the model at all. The *subject-specific* estimates, or BLUPs, are found in the Random Coefficients section of the report. The terms "population-average" and "subject-specific" come from Zeger et al. (1988). Remember, these are the estimates of the b_{0i} and b_{1i} . To find complete batch-specific intercept or slope, add the overall estimate with the batch-specific estimate.

Figure 5.2: Fit Mixed Report for the Stability Trial Data

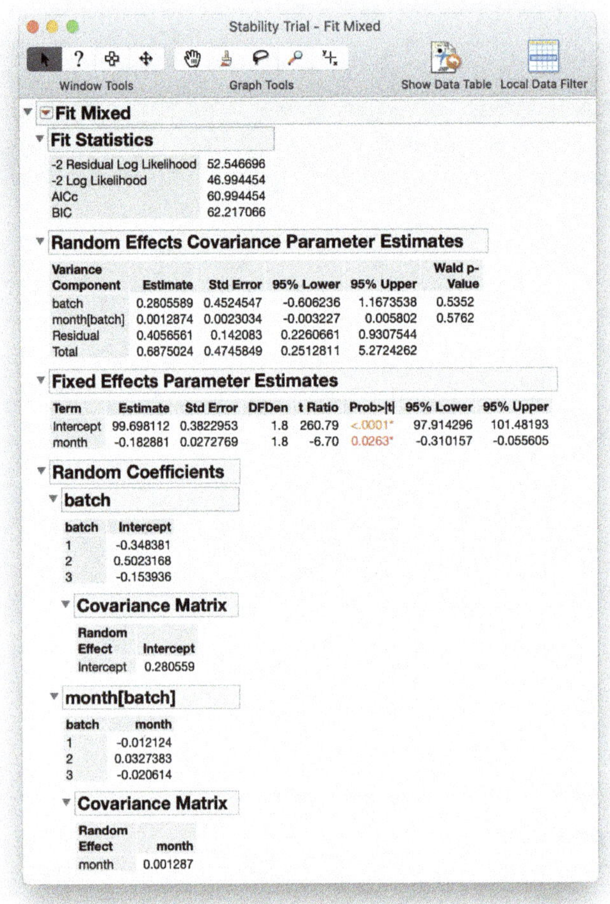

The goal of the study was to find the shelf life for the drug being tested. It is typically mandated that the probability of all future batches lasting at least as long as the stated shelf life is 95%. We can do a 95% one-sided interval around the "worst" performing batch to estimate this. Stroup and Quinlan (2016) provide evidence that this lower bound provides an estimate that is both accurate and addresses the "all future batches" mandate.

We obtain the interval using the Inverse Prediction option in the red triangle menu.

Figure 5.3 shows the completed inverse prediction dialog box. We do not know which batch is the worst performing, so check the box for *All* batches to predict. The confidence level is what we want, but we need to change from *Two sided* to *Lower One Sided* in the drop-down box. The Y value of interest is 90. Finally, select the check box for "Confidence interval with respect to individual rather than expected response" because we are predicting an individual value and not the mean. This has greater uncertainty and will result in a wider interval.

Figure 5.3: Fit Mixed Report for the Stability Trial Data

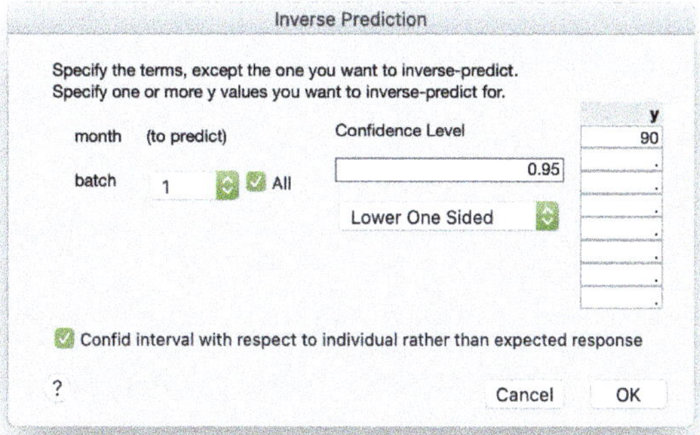

After pressing *OK*, the results of the inverse prediction are added to the Fit Mixed report shown in Figure 5.4. Of the three batches, batch 3 has the smallest predicted month, 46.9, for reaching the level of the stability limiting characteristic indicating it is the worst performing batch. The lower limit for the confidence interval is 38.47 months, so the shelf life for this product would be set at 38 months. We reiterate here that this procedure is for products that do not have regulatory directives for how to set shelf life.

Figure 5.4: Fit Mixed Report for the Stability Trial Data

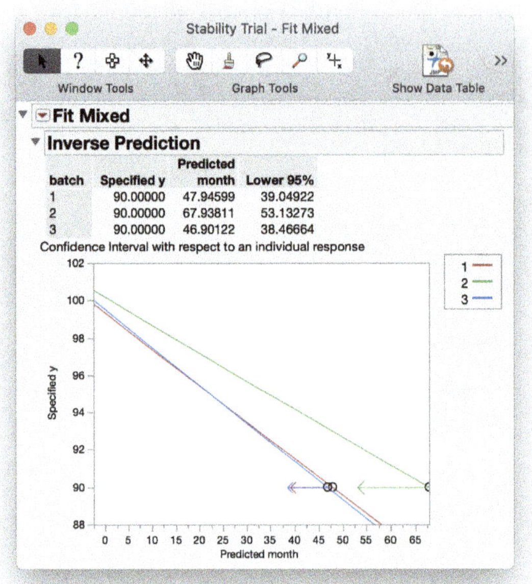

5.4 Student Achievement Example

The social sciences including education research have long used Hierarchical Linear Models (HLMs) due to the structure inherent in their data, for example, counties within states within countries, or students within classrooms within schools. Singer (1998) made the translation from the notation and terminology of the HLM to the random coefficient model demonstrating how these models could be fit in SAS PROC MIXED. In this spirit, we will make this same translation from the hierarchical framework to the form required for fitting in JMP.

The `SIMS.jmp` data table contains a subset of the results of the Second International Mathematics Study (SIMS), which measured eighth-grade students on mathematics achievement tests. It was analyzed in Kreft et al. (1994). The 3,691 students are grouped into 190 different classes yielding the hierarchical structure. The students are the level at which we measure the response, thus the lowest level or level one. The classes are at the next higher level, level two. There is a student-level predictor, `pretot`, a pretest score for each student, and the response variable, `gain`, the change in score from the pretest. The goal is to examine academic achievement gains as a function of the pretest scores.

HLM to Random Coefficient Model

In HLM terms the level 1 model is

$$gain_{ij} = \beta_{0j} + \beta_{1j}(pretot)_{ij} + e_{ij}$$

$gain_{ij}$ is the gain on the score of the achievement test by student i in the jth class. $pretot_{ij}$ is the pretest score for the same student. The errors, e_{ij}, are independent and identically distributed with mean zero and variance σ^2.

The regression coefficients, β_{0j} and β_{1j} are assumed to arise at the HLM level 2, classroom level. The level two model is

$$\beta_{0j} = \gamma_{00} + u_{0j}$$
$$\beta_{1j} = \gamma_{10} + u_{1j}$$

where

$$\begin{bmatrix} u_{0i} \\ u_{1i} \end{bmatrix} \sim N\left(\begin{bmatrix} 0 \\ 0 \end{bmatrix}, \begin{bmatrix} \sigma_{00} & \sigma_{01} \\ \sigma_{01} & \sigma_{11} \end{bmatrix}\right)$$

γ_{00} is the intercept, and the u_{0j} are the class-level random deviations around the intercept.
γ_{10} is the population average slope, while the u_{1j} are the deviations around the overall slope.
The γ terms are the fixed effects in the model, while the u terms are the random effects.
We assume that the u_{0j} and u_{1j} are independent, identically distributed as bivariate normal with the variance-covariance matrix as shown.

To "flatten" the two levels of the hierarchy into the statistical modeling form we use in this book, we can substitute the level 2 equations into the level 1 equation for one overall modeling equation. When we do, we can see that the HLM is another way of expressing the same random coefficient model.

$$gain_{ij} = (\gamma_{00} + u_{0j}) + (\gamma_{10} + u_{1j})(pretot)_{ij} + e_{ij}$$

This method of building the full equation can be extended to arbitrarily complex models where there could be more than two levels or predictors at any of the levels. Then fitting the model in JMP is straightforward.

Random Coefficient Instructions for Mathematics Achievement Data

With the `SIMS.jmp` data table open, go to *Analyze > Fit Model*. Enter `gain` as the *Y* variable. Change the default personality from *Standard Least Squares* to *Mixed Model*.
Select `pretot` in the *Select Columns* box then click *Add* to add fixed effect of the pretest score to the *Construct Model Effects* box.
Enter `pretot` in the *Random Effects* tab of the *Construct Model Effects* box. Then select `class` in the *Select Columns* box and `pretot` in the *Random Effects* tab and click *Nest Random Coefficients* to create the class-specific random intercept and slope.
Click *Run* to fit the model.

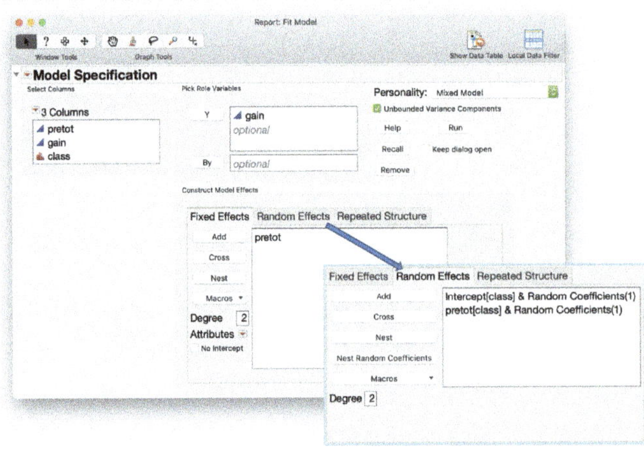

Results and Interpretation

The fits shown in Figure 5.5 look reasonable for this type of data. There are many factors that can influence a student's score on a math test, not just how they performed on a pretest. With only the pretest score factor, a "perfect" fit is not possible, so we would expect some larger deviations away from the 45-degree line. In the Fixed Effects Parameter Estimates, the Intercept of 7.05 means that for an average student with a pretest score of 0, the gain in their score would be about 7. The negative slope estimate of -0.186 for `pretot` is perhaps counterintuitive. It means that the higher the pretest score, the lower the amount of gain the student achieves on average. Thinking further, that makes some sense in that there is likely a ceiling effect; a student can only score so high. If they already scored high on the pretest, they have less room to grow.

The ceiling effect is also shown in the negative covariance between the intercept and pretot slope. This means that classrooms with higher intercepts had lower slopes, and those with lower intercepts had higher slopes. If a class starts higher, the decline is greater than for a class that starts lower on average. The confidence interval for the covariance does not cross zero, so we can be confident that this covariance is important to the model.

Figure 5.5: Fit Mixed Report for the Mathematics Achievement Data

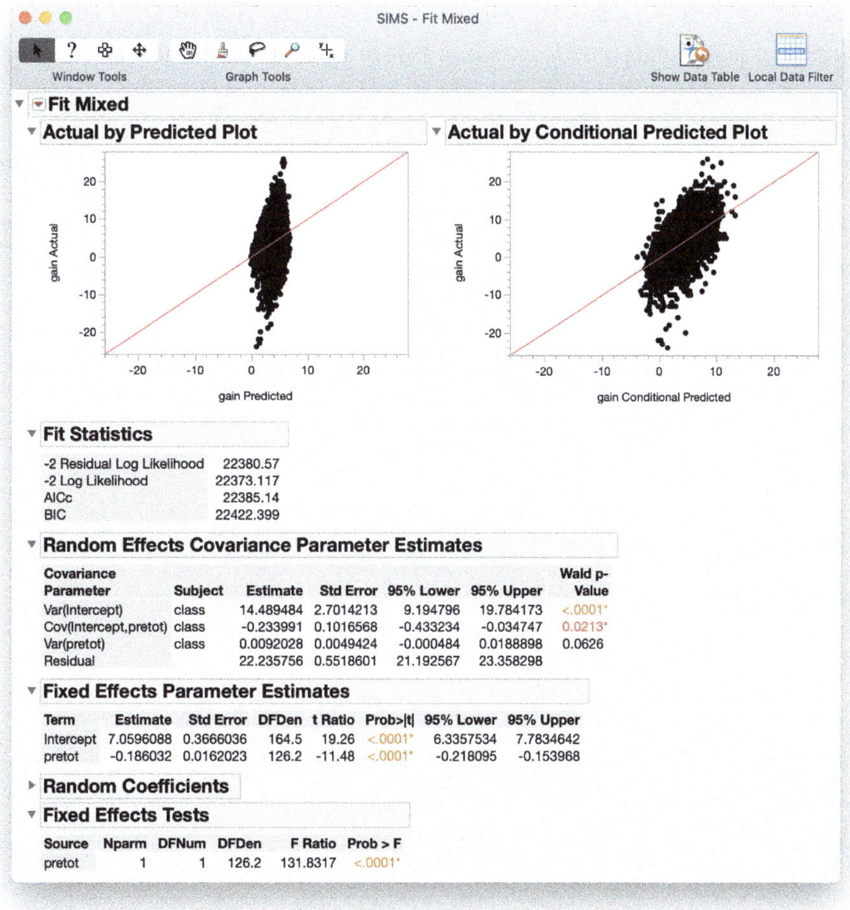

Finally, Kreft et al. (1994) also analyzed the data with a class-level predictor, `OTL` for Opportunity to Learn, measured at each classroom. Unfortunately, the data with this variable have been lost to the mists of time, so we cannot fully analyze it here. We will show how the level 2 equations change to include this predictor, the resulting full model equation, and the model dialog for this model if we had the data.

HLM with level 2 Predictor to Random Coefficient Model

The class-level model is modified from before as follows.

$$\beta_{0j} = \gamma_{00} + \gamma_{01}(OTL)_j + u_{0j}$$

$$\beta_{1j} = \gamma_{10} + \gamma_{11}(OTL)_j + u_{1j}$$

where γ_{01} is the fixed slope for `OTL` and γ_{11} is the fixed effect for the interaction of `OTL` and `pretot`. This interaction will be clear below, because it is not obvious from these equations. The assumptions about the u are the same.

These augmented level two equations are then substituted back into the level one model for a final model.

$$gain_{ij} = (\gamma_{00} + \gamma_{01}(OTL)_j + u_{0j}) + (\gamma_{10} + \gamma_{11}(OTL)_j + u_{1j})(pretot)_{ij} + e_{ij}$$

We can rearrange terms to put all of the fixed effects first followed by the random effects for clarity.

$$gain_{ij} = \gamma_{00} + \gamma_{01}(OTL)_j + \gamma_{10}(pretot)_{ij} + \gamma_{11}((OTL)_j(pretot)_{ij}) + u_{0j} + u_{1j}(pretot)_{ij} + e_{ij}$$

It is now clear that besides the main effects of `OTL` and `pretot` we are including an effect for the possible interaction between them.

Expanded Model Instructions for Mathematics Achievement Data

With the **SIMS.jmp** data table open, go to *Analyze > Fit Model*. Enter **gain** as the *Y* variable. Change the default personality from *Standard Least Squares* to *Mixed Model*.

Select both **pretot** and **OTL** using CNTL-click in the *Select Columns* box then click *Macro > Full Factorial* to add fixed effect of the pretest score, the OTL measurement and their interaction to the *Construct Model Effects* box.

Enter **pretot** in the *Random Effects* tab of the *Construct Model Effects* box. Then select **class** in the *Select Columns* box and **pretot** in the *Random Effects* tab and click *Nest Random Coefficients* to create the class-specific random intercept and slope. This part of the model is the same as before and is not shown below.

Click *Run* to fit the model.

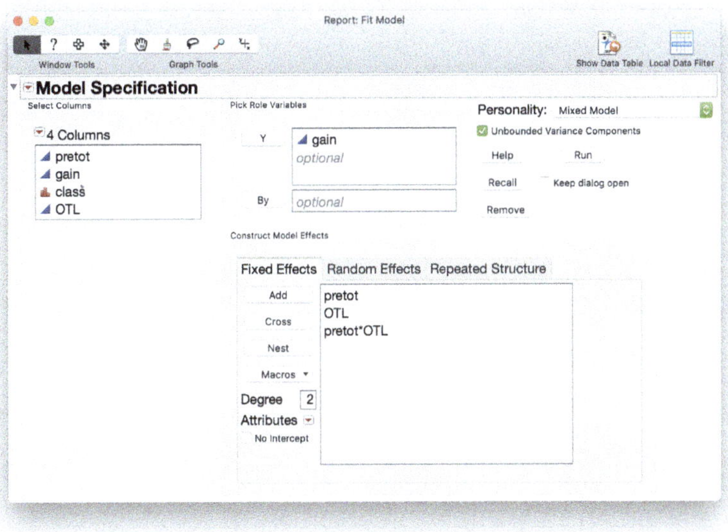

5.5 Exercises

1. The data table **winterwheat.jmp** contains data from ten varieties of wheat that were randomly selected from the population of varieties of hard red winter wheat. The varieties were randomized to sixty, 1-acre plots in a field for six replicates for each variety. It was thought that the preplanting moisture of the soil could influence the germination rate and hence the eventual yield. So, the amount of moisture in the top 36 inches of soil was determined from a core sample at the middle of each plot. The response is bushels per acre, **yield**, and the covariate is the amount of preplanting moisture, **moist**.

Is a random coefficient model appropriate for these data? If so, why? If not, why not?

2. Regardless of how you answered question 1, fit a random coefficient model for these data in JMP. Inspect the fit. Does it look reasonable? Is the random coefficient model necessary or can it be simplified?
3. Either with the random coefficient model or a simplified model as you deem appropriate, does there appear to be a better performing variety of wheat? Explain with evidence.

Chapter 6

Repeated Measures and Longitudinal Data

In this chapter, we examine experiment designs that require repeatedly measuring the same experimental units, whether by taking many different types of measurements on the same unit or taking repeated measurements of the same type, over time, on the same unit. When you take repeated measurements, these repeated observations are no longer independent of each other.

With repeated measures analyses, we will still answer the standard mixed model questions about which treatment and experiment design factors demonstrate statistically significant relationships to the response and where those differences lie. However, the special feature in this type of repeated measures analysis is that we must include an assumption, and the related model-fitting, about how the repeated measures are dependently related to each other.

6.1 Motivating Example

Respiratory Ability — A pharmaceutical company examines effects of three drugs on respiratory ability of asthma patients. Treatments include a standard drug (A), a test drug (C), and a placebo (P). The drugs are randomly assigned to 24 patients each. The assigned treatment is administered to each patient, and a standard measure of respiratory ability called FEV1 is measured hourly for 8 hours following treatment. FEV1 is also measured immediately prior to administration of the drugs.

6.2 Conceptual Background

Two popular historical solutions to the problem of multiple responses, the MANOVA and the Split-Plot in Time, are often insufficient in addressing the true correlation structure. We introduce these, along with some weaknesses therein, and a more adaptable strategy to directly model the correlation and covariance introduced from repeated measures.

MANOVA

One solution to the idea of taking multiple measurements on each subject is the multiple-response ANOVA (MANOVA). MANOVA allows the correlations between the response variables within subjects to be estimated individually and incorporated into an ANOVA model with multiple responses. The MANOVA is actually, conceptually, a special case of one of the mixed models structures that we will present in this chapter. The MANOVA corresponds to the model with the fewest restrictions – the Unstructured model.

MANOVA is often used when sets of different measurement types are taken on the same subject (rather than repeatedly taking the same type of measurement, over time). However, MANOVA (or, equivalently, the Unstructured model that we will explore in the following section) tends to have low power, especially as the number of response variables or time points increases (Crowder and Hand, 1990). When we have many repeated measurements of the same type over time, MANOVA is not the most efficient way to use our repeated information. Instead, we would like to determine how the repeated measurements within subjects are related to each other, and account for this directly in a model that includes the repetition as an effect and predicts only the single response variable.

We include some additional information about MANOVA and addressing multiple unrelated responses taken on the same subjects in Section 6.5.

Split-Plot in Time

Measurements taken on the same experimental unit are correlated in a manner similar to blocking. In fact, a very simple way to address these repeated measurements is by treating the common Subject as a block, as in the Compound Symmetry model.

This is the idea of the traditional Split-Plot in Time analysis for addressing repeated measures. The issue with this approach is that it is often not appropriate or well-fitting to force all of the time points to be equally correlated with each other. In repeated measurements experiments, it is far more often the case that measurements taken nearer in time are more highly correlated. For this reason we will allow this model to be included in our suite of models to consider, but we expect that it will not generally fit well to real repeated measures data.

This is the only repeated measures structure available in JMP. For this reason, most of this chapter focuses on options and outputs in JMP Pro, which allows much more flexibility for fitting repeated measures models.

Correlated within Subject

We specify a repeated measures correlation structure to account for a lack of independence among observations. When repeated measurements are taken on the same subject, we expect those observations to be correlated. To capture this correlation, we need to fit a reasonable set of assumptions to it. The idea is that you tell the model, for example, "okay, these observations are not all independent. In fact, within any subject, the time 1 measurement will be correlated with the time 2 measurement with a correlation of 0.8. And the time 2 measurement will be correlated with time 3 with a correlation of 0.8. In fact, any measurements (for the same subject) that are 1 time period apart will be correlated with 0.8. Then for times that are 2 periods apart, the correlation will drop to $0.8^2 = 0.64$. And times that are three periods apart will drop to $0.8^3 = 0.51$. And so on." (This is a description of the AR(1) structure, with $r = 0.8$, which we will explore further later in this section.)

Instead of claiming that the observations are all independent of each other, we claim that the observations are all independent from each other *after accounting for correlations within the measurements of a single subject over time.*

We are saying that the first observation on a subject is totally random, generated from a normal distribution with a certain mean and variance. After fitting the treatment model, let's say that first observation, at Time 1, for Subject 1, has a residual (compared to its model-predicted value) of 4.5. Now the observation for Subject 1 at Time 2 is going to have an expected residual of $0.8 * 4.5 = 3.6$, with the same variance as Subject 1 Time 1 had. Maybe this Subject 1 Time 2 residual is actually $3.94 = 3.6 + 0.34$, where the 0.34 is the randomly generated and independent component. This is the claim of forcing a specific correlation structure: we are telling the model that a certain amount of similarity between observations within the same subject can be explained, and the rest is random and independent variation.

This type of model, where we specify a correlation structure within subjects over time, does not change how the Mixed Model estimates the means for each Subject or Time group, but it does change the variance estimates for comparisons between levels of those Subject or Time groups. You may find that the statistical significance of comparisons changes as you change the correlation structure assumption in the model. For example, perhaps Time 1 and Time 4 have a mean difference of -1.2 and this difference is statistically significant when you do not specify a correlation structure, but it is not statistically significant when you specify an AR(1) correlation structure. In such a case, you might falsely find a statistically significant difference that does not really exist once the appropriate assumptions are made.

This is the purpose for modeling the correlation and covariance structures for repeated measures models — to propose a model that most closely matches reasonable assumptions of dependence between the repeated measures and thereby correct the variance estimates in your hypothesis tests and confidence interval estimates.

Modeling the Covariance Structure

By modeling the covariance structure, we are attempting to fit, and then account for, the relationship of dependence within the subjects and over time. As with straightforward random effects, these variances and covariances are then used to help inform the confidence intervals and tests for the factors of interest.

We have already noted that the Unstructured assumption (equivalent to the MANOVA) is the fullest, most parameterized, least restrictive in our set of possible structures. The

Compound Symmetry structure is the simplest, least parameterized, most restrictive in our set of structures. We will see a few more options that fall within this spectrum as well.

Let's begin by examining the relationships between time points by considering their Variance-and-Covariance and Correlation matrices. We will always begin by examining the *Unstructured* repeated covariance option. The matrices for an example with four repeated measurements would be diagonal 4-by-4 matrices, with each column corresponding to the first, then second, then third, then fourth time points, and the rows corresponding to the same. We read the variances off of the diagonal of the Variance-Covariance matrix, and we read the covariances from the off-diagonals. For example, row 2 column 3 is the covariance of the response at time 2 with the response at time 3. In the Correlation Matrix, the diagonals will always be 1. The off-diagonals correspond to the correlations in the response between time points.

Unstructured Var/Covariance Matrix

var_1	$cov_{1,2}$	$cov_{1,3}$	$cov_{1,4}$
	var_2	$cov_{2,3}$	$cov_{2,4}$
		var_3	$cov_{3,4}$
			var_4

Unstructured Correlation Matrix

1	$r_{1,2}$	$r_{1,3}$	$r_{1,4}$
	1	$r_{2,3}$	$r_{2,4}$
		1	$r_{3,4}$
			1

For example:

Unstructured Var/Covariance Matrix

12.86	1.01	-2.08	2.39
	10.07	-0.63	-3.21
		7.16	-2.04
			6.32

Unstructured Correlation Matrix

1	0.09	-0.22	0.26
	1	-0.07	-0.40
		1	-0.30
			1

In this example, all of the individual cell values are different from each other. But we still want to explore these to see whether there might be a pattern.

There are two things to look for:

1. Do you have the same variances down the diagonal of the covariance matrix? (We must decide whether we will assume equal versus unequal variances for each time point.)
2. Is there a pattern to the correlations as you move from time lag 1 to longer time lags? (We must decide which patterns – or no pattern – to try in our model.)

We will examine these structure options later in this section:

- Correlation is the same everywhere (CS)
- Correlation decays over time gap size (AR(1) or spatial)
- Correlation differs over time gap size, but without a pattern (Toeplitz)

- Correlation differs at lag 1 gaps, but hold a pattern over time (Antedependent)

Step 1: Are the variances equal or unequal?

The first pattern to look for is: are the variances equal or unequal? Just look down the diagonal. A common rule-of-thumb is to use an equal variances assumption if all of the variances are within four-fold of each other and to use an unequal variances assumption if any of the variances are greater than four times another. (This rule-of-thumb translates to a two-fold difference in standard deviations.) One important note: In JMP, the `Unequal Variances` structure assumes unequal variances but does not allow any nonzero correlations. This is not actually a repeated measures structure. It is equivalent to including Time as a random effect to account for the variation at each time point.

Step 2: Which correlation structures might fit the Unstructured correlation matrix?

After deciding on equal or unequal variances, next look at the Unstructured correlation matrix. Compare this to the possible structures available in JMP: Compound Symmetry, Toeplitz, and Antedependent, all of which can be modeled with equal or unequal variances, and Auto-Regressive-of-Lag-1 (AR(1)) and the spatial options. Choose the potential contenders, fit each of these and compare the resulting fit statistics. (More about this at the end of this section.)

Let's look more closely at each repeated measures correlation and covariance structure option.

Compound Symmetry

Compound Symmetry, as we mentioned in Section 6.2, is identical to a model with a random effect for the Subject and no other modeling of the correlation over time. This structure will fit well when the correlations between time points remain constant over any time lag.

Compound Symmetry Correlation Matrix

1	r	r	r
	1	r	r
		1	r
			1

Example CS Correlation Matrix

1	.97	.97	.97
	1	.97	.97
		1	.97
			1

AR(1)

Auto-Regressive-of-Order-1 (AR(1)) holds the correlations constant for observations at any two time points of lag 1, and then allows that correlation to decay exponentially as the time lag increases. Many statistical software packages require the time points to be at equal intervals, but JMP allows unequal spacing in the time points.

Auto-Regressive(1) Correlation Matrix

1	r	r^2	r^3
	1	r	r^2
		1	r
			1

Example AR(1) Correlation Matrix

1	.90	.81	.73
	1	.90	.81
		1	.90
			1

Spatial

We will cover the spatial structure more in Chapter 7, but this structure, used with Time as the Repeated factor, can also be used to capture an AR(1) or exponential trend where the "spatial" element is just a continuous time measurement. In fact, the Spatial Power trend without a nugget is identical to the AR(1) model. We will cover this more in the next chapter, when addressing spatial models for distance rather than time.

Spatial Correlation Matrix

1	r_1	r_2	r_3
	1	r_1	r_2
		1	r_1
			1

Example SpPow Correlation Matrix

1	.80	.64	.51
	1	.80	.64
		1	.80
			1

Toeplitz

The Toeplitz pattern has one correlation for all of the lag 1 cells in the correlation matrix, a different correlation, unrelated to the lag 1 correlation, for all of the lag 2 cells, and so on. Each diagonal lag band has the same correlation throughout, and there is no trend from one band to the next.

Toeplitz Correlation Matrix

1	r_1	r_2	r_3
	1	r_1	r_2
		1	r_1
			1

Example Toep Correlation Matrix

1	.70	.95	-.60
	1	.70	.95
		1	.70
			1

Antedependent

The Antedependent pattern is harder to see from the matrices. Use this when the lag 1 correlations are dissimilar (unlike Toeplitz or AR(1)) but you still have a pattern over time (like AR(1)). This structure also works well for unequal spacing for the time measurements.

Antedependent Correlation Matrix

1	r_{12}	$r_{12}r_{23}$	$r_{12}r_{23}r_{34}$
	1	r_{23}	$r_{23}r_{34}$
		1	r_{34}
			1

Example Ante Correlation Matrix

1	.70	-.28	-.16
	1	-.40	-.22
		1	.56
			1

Summary of Repeated Structures Available in JMP Pro

Table 6.2, adapted from the JMP Help Documentation on Fitting Linear Models (SAS Institute Inc. 2019a), shows the repeated measures structure types available in JMP Pro and details on the required repeated measurement data type. The repeated measurement data type can be *categorical* (with an alphanumeric ordering) to treat the time points as groups or *continuous* to treat the time points as points in a continuous timeline. For structures that accept continuous time, these times do not need to be equally spaced. For structures that accept categorical time, the model will generally assume equal spacing, although structures that allow different estimates over time (such as the Antedependent) can be interpreted without requiring that assumption.

Table 6.1: Repeated Covariance Structures Available in JMP Pro; Number of Measurement Times Is Denoted by J

Structure	**Repeated Column Type**	**Number of Params**	**Details**
Unequal Variances	categorical	J	Not a true repeated measures model (equivalent to using time as a random effect) – don't use this
Unstructured	categorical	$J(J+1)/2$	Fullest possible model – fit this first and look for trends
Compound Symmetry	categorical	2	Split-Plot in Time; requires constant correlations at any time lag
Compound Symmetry Unequal Variance	categorical	$J+1$	Split-Plot in Time; requires constant correlations at any time lag, plus unequal variances across time points
Antedependent (called "Antedependent Equal Variances")	categorical	J	Allows dissimilar correlations within a lag but retains a pattern over time
Antedependent Unequal Variance (called "Antedependent")	categorical	$2J-1$	Allows dissimilar correlations within a lag but retains a pattern over time, plus unequal variances across time points
Toeplitz	categorical	J	Allows different correlations for each time lag
Toeplitz Unequal Variance	categorical	$2J-1$	Allows different correlations for each time lag, plus unequal variances across time points
AR(1)	continuous	2	Allows an auto-regressive trend in the correlations between measurements within a subject over time
Spatial(Power)	continuous	2	Identical to AR(1)
Spatial(Exponential)	continuous	2	Similar to AR(1) but with exponential trend instead of power
Spatial with nugget	continuous	2	Allows an intercept-like term for the covariance at an infinitesimally small distance in time

Step 3: Fitting the possible structures, comparing fit statistics, and choosing the best fit

The final step in this modeling process is to fit the full planned model (treatment and experiment design elements) using each of the identified candidate structures. Examine the resulting fit statistics for each model and select a model that fits well. Once this structure has been chosen, then interpret the treatment and experiment factor tests using the model with this best-fitting underlying correlation and covariance structure.

6.3 Repeated Measures Skeleton ANOVA and Statistical Model

A possible Skeleton ANOVA for a scenario with a Treatment assigned to each group of Subjects, and repeated measurements taken over time on the Subjects, follows. The Treatment could be a single factor or could be replaced by a factorial combination of treatment factors. This version of the Skeleton ANOVA is specific to the model with the

Experiment Design	Skeleton ANOVA
Source	**Source**
	Treatment
Subject(Treatment) (like Whole-plot e.u.)	Between subjects
	Time
	Treatment*Time
Measurements(Subject)	Within subjects
Total	Total

Compound Symmetry structure, i.e., the split-plot in time model of Section 6.2. Compare this to the Skeleton ANOVA in Chapter 3 to see the equivalence. When attempting to write the Skeleton ANOVA and the corresponding model for any more complex covariance structures, the Between Subjects and Within Subjects terms will become too complicated to express in a simple Skeleton ANOVA. We present a general statistical model in the following section.

The Statistical Model for Repeated Measures with Unstructured, Compound Symmetry, AR(1), Toeplitz, or Antedependent Structure

The model specification changes depending on the type of correlation structure we want to impose. In general, we can write the statistical model as follows, but the line about the Subject effects, b_{ij} and w_{ijk}, captured in the e_{ijk} term, will change to match the specified structure, as described in Littell et al. (2000).

Model for One Treatment Factor with Repeated Measures

$$y_{ij(k)} = \mu + \alpha_i + b_{ij} + \gamma_k + (\alpha\gamma)_{ik} + w_{ijk}$$

y_{ijk} is the response at the kth time on the jth subject in the ith treatment.
μ is the intercept.
α_i is the effect of the ith treatment.
b_{ij} is the random *between-subjects* effect of the ij^{th} subject.
γ_k is the effect of the kth time.
$\alpha\gamma_{ik}$ is the interaction effect of the kth time with the ith treatment.
w_{ijk} is the random *within-subjects* effect of the kth time on the ij^{th} subject.
If we combine the between- and within-subjects random effects, we have the typical residual: e_{ijk} is the residual error and $e_{ijk} = b_{ij} + w_{ijk}$, where errors for different subjects are independent and $Cov(e_{ijk}, e_{i'j'k})$ is determined by the specified structure. for $i \neq i'$ or $j \neq j'$.

6.4 Respiratory Ability

This example appears as Data Set 5.2, `Respiratory Ability`, in Appendix 2 of Littell et al. (2006). It originally appeared in Littell et al. (2000) and was used in Littell et al. (2002).

Figure 6.1: Respiratory Ability Data Stacked, All Responses in a Single Long Column

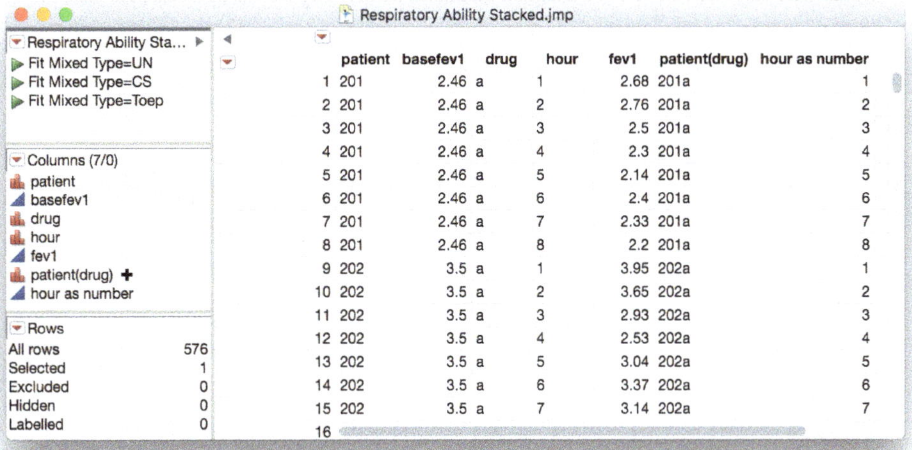

Respiratory Ability Stacked.jmp

Respiratory Ability Sta...
Fit Mixed Type=UN
Fit Mixed Type=CS
Fit Mixed Type=Toep

Columns (7/0)
patient
basefev1
drug
hour
fev1
patient(drug)
hour as number

Rows
All rows 576
Selected 1
Excluded 0
Hidden 0
Labelled 0

	patient	basefev1	drug	hour	fev1	patient(drug)	hour as number
1	201	2.46	a	1	2.68	201a	1
2	201	2.46	a	2	2.76	201a	2
3	201	2.46	a	3	2.5	201a	3
4	201	2.46	a	4	2.3	201a	4
5	201	2.46	a	5	2.14	201a	5
6	201	2.46	a	6	2.4	201a	6
7	201	2.46	a	7	2.33	201a	7
8	201	2.46	a	8	2.2	201a	8
9	202	3.5	a	1	3.95	202a	1
10	202	3.5	a	2	3.65	202a	2
11	202	3.5	a	3	2.93	202a	3
12	202	3.5	a	4	2.53	202a	4
13	202	3.5	a	5	3.04	202a	5
14	202	3.5	a	6	3.37	202a	6
15	202	3.5	a	7	3.14	202a	7
16							

A pharmaceutical company examined effects of three drugs on respiratory ability of asthma patients. Treatments were a standard drug (A), a test drug (C), and a placebo (P). The drugs were randomly assigned to 24 patients each. The assigned treatment was administered to each patient, and a standard measure of respiratory ability called FEV1 was measured hourly for 8 hours following treatment. FEV1 was also measured immediately prior to administration of the drugs.

The original data table is in the wide format, with different columns for each time point. (One row per patient, but the response is in eight different columns, according to the time point.) To more easily analyze the time trend, it is useful to first create a stacked version of the data, where each patient now has eight rows and the response FEV1 is now in a single column. It is also helpful to create a column **`hour as number`** to allow fitting covariance structures like AR(1), spatial, and Antedependent that require a continuous measurement of time. (More about this later.) Lastly, it is helpful to create a column called **`patient(drug)`** to uniquely identify each patient in a single column. (The patient numbers repeat for the three drug groups, but patient 201 in Drug C is NOT the same person as patient 201 in Drug P.) This stacked version of the data, with the additional columns, is shown in Figure 6.2 and is in a file called **`Respiratory Ability Stacked.jmp`**.

Patient Trends over Time

Using Graph Builder in JMP, we can view the trends of the patients within each drug group over time. Figure 10.2 shows the patient profiles. The Placebo group shows some random variation over time but no general trend. Drugs C and A both appear to have downward trends in FEV1 over time, with Drug C slightly higher on average across time than Drug A.

Figure 6.2: Patient FEV1 Measurements over Time

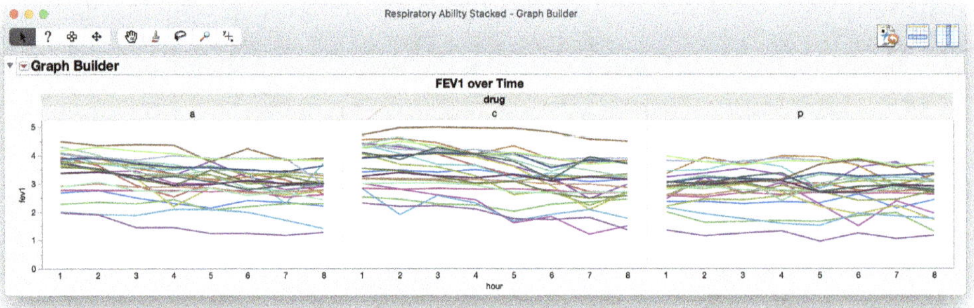

We can also write out the Skeleton ANOVA and the Statistical Model. Because the Drug groups (part of the Treatment Design) are also the Blocks (called an *arm* in clinical drug trials language) in which the Patients are nested (part of the Experiment Design), we can include this Arm=Drug information in both the Treatment and the Experiment columns. However, since these two terms are really identifying the same thing, we only pull one of them over into the final column to describe the full design.

Experiment Design		**Skeleton ANOVA**	
Source	***df***	**Source**	***df***
Arm	3-1	Drug	2
Patient(Arm)	3 x (24-1) = 69	Between Subjects	69
		Hour	8-1 = 7
		Drug*Hour	2*7 = 14
Obs(Patient(Arm))	(3*24)*(8-1) = 504	Within Patients	3*(24-1)*(8-1) = 483
Total	(3 x 24 x 8) - 1 = 575	Total	575

Model for Respiratory Ability

$$y_{ijk} = \mu + \gamma x_{ij} + \alpha_i + \tau_j + (\alpha\tau)_{ik} + e_{ijk}$$

y_{ijk} is the response of the kth hour measurement on the jth patient in the ith drug group.
μ is the intercept.
γx_{ij} is the fixed effect of the baseline FEV1 measurement for the jth patient in the ith drug group.
α_i is the effect of the ith drug group.
τ_k is the effect of the kth hour.
$\alpha\tau_{ik}$ is the interaction effect of the ith drug group with the kth hour.
e_{ijk} is the ijk^{th} error, where errors for different subjects are independent and the vector of errors is ~ $N(0,R)$ with the specified covariance structure R.

JMP Instructions for Respiratory Ability – CS Structure Only

Because the Compound Symmetry structure is equivalent to specifying the Subject as a random blocking variable, we can use standard JMP to fit a repeated measures analysis for the Compound Symmetry structure. Simply specify the Subject as a random effect in the Model Effects box. Compound Symmetry is the only repeated measures structure that we can fit using standard JMP; all other specialized structures require JMP Pro. For this reason, we will focus on the output from JMP Pro for this example.

JMP Pro Instructions for Respiratory Ability

Using JMP Pro

Use *Analyze > Fit Model* to define the model. Change the Personality to *Mixed Model* in order to see the tabs for *Construct Model Effects* that include *Random Effects* and *Repeated Structure*.

Enter `FEV1` as the *Y* variable. Enter `drug`, `hour`, and `drug*hour` as *Fixed Effects*. (You could also put `basefev1` here if you want to adjust each subject by their baseline FEV1 measurement.) You do not need to enter any *Random Effects* in this case, but in a different example, you may have other random effects to add here. For the *Repeated Structure*, select *Unstructured* and enter `hour` as the *Repeated* and `patient(drug)` as the *Subject*.

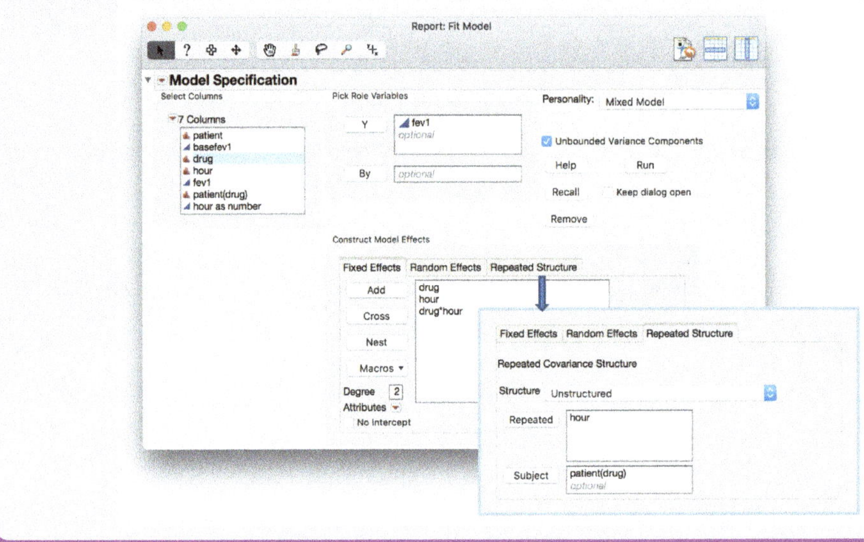

Results and Interpretation

The reported variance/covariance information is shown in Figure 6.3.

Looking at the Covariances and Correlations to Choose a Structure

To make it easier to visualize the variances and covariances from the output matrix, we can use a JMP add-in designed for this. Add-ins are extensions to the base functionality of JMP, and JMP users can write and distribute and use these JMP extensions to add point-and-click functionality to JMP. Download the **Repeated Measures Cov/Corr Diagnostics** add-in file (extension ".jmpaddin") from `community.jmp.com` and click on the file to open it (Parris and Hummel, 2021b). You will be prompted to install the add-in. Click the *Install* button. You can now delete the original file. The add-in is permanently

Figure 6.3: Unstructured Estimates of Variance and Covariance

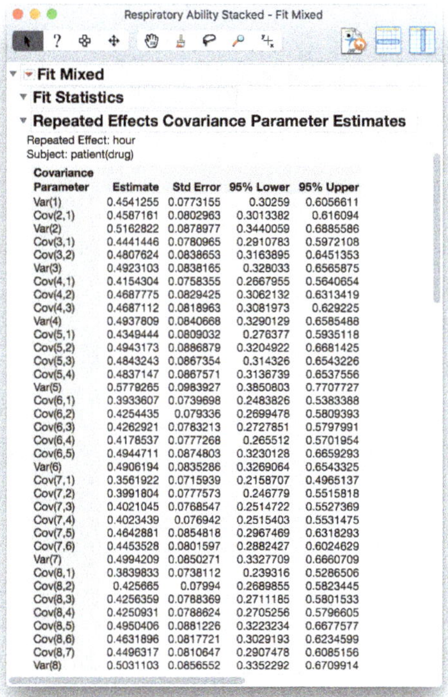

Respiratory Ability Stacked - Fit Mixed

Fit Mixed

Fit Statistics

Repeated Effects Covariance Parameter Estimates

Repeated Effect: hour
Subject: patient(drug)

Covariance Parameter	Estimate	Std Error	95% Lower	95% Upper
Var(1)	0.4541255	0.0773155	0.30259	0.6056611
Cov(2,1)	0.4587161	0.0802963	0.3013382	0.616094
Var(2)	0.5162822	0.0878977	0.3440059	0.6885586
Cov(3,1)	0.4441446	0.0780965	0.2910783	0.5972108
Cov(3,2)	0.4807624	0.0838653	0.3163895	0.6451353
Var(3)	0.4923103	0.0838165	0.328033	0.6565875
Cov(4,1)	0.4154304	0.0758355	0.2667955	0.5640654
Cov(4,2)	0.4687775	0.0829425	0.3062132	0.6313419
Cov(4,3)	0.4687112	0.0818963	0.3081973	0.629225
Var(4)	0.4937809	0.0840668	0.3290129	0.6585488
Cov(5,1)	0.4349444	0.0809032	0.276377	0.5935118
Cov(5,2)	0.4943173	0.0886879	0.3204922	0.6681425
Cov(5,3)	0.4843243	0.0867354	0.314326	0.6543226
Cov(5,4)	0.4837147	0.0867571	0.3136739	0.6537556
Var(5)	0.5779265	0.0983927	0.3850803	0.7707727
Cov(6,1)	0.3933607	0.0739698	0.2483826	0.5383388
Cov(6,2)	0.4254435	0.079336	0.2699478	0.5809393
Cov(6,3)	0.4262921	0.0783213	0.2727851	0.5797991
Cov(6,4)	0.4178537	0.0777268	0.265512	0.5701954
Cov(6,5)	0.4944711	0.0874803	0.3230128	0.6659293
Var(6)	0.4906194	0.0835286	0.3269064	0.6543325
Cov(7,1)	0.3561922	0.0715939	0.2158707	0.4965137
Cov(7,2)	0.3991804	0.0777573	0.246779	0.5515818
Cov(7,3)	0.4021045	0.0768547	0.2514722	0.5527369
Cov(7,4)	0.4023439	0.076942	0.2515403	0.5531475
Cov(7,5)	0.4642881	0.0854818	0.2967469	0.6318293
Cov(7,6)	0.4453528	0.0801597	0.2882427	0.6024629
Var(7)	0.4994209	0.0850271	0.3327709	0.6660709
Cov(8,1)	0.3839833	0.0738112	0.239316	0.5286506
Cov(8,2)	0.425665	0.07994	0.2689855	0.5823445
Cov(8,3)	0.4256359	0.0788369	0.2711185	0.5801533
Cov(8,4)	0.4250931	0.0788624	0.2705256	0.5796605
Cov(8,5)	0.4950406	0.0881226	0.3223234	0.6677577
Cov(8,6)	0.4631896	0.0817721	0.3029193	0.6234599
Cov(8,7)	0.4496317	0.0810647	0.2907478	0.6085156
Var(8)	0.5031103	0.0856552	0.3352292	0.6709914

installed in your *Add-Ins* menu in JMP. This add-in will take the unstructured repeated measures covariance and correlation matrices and report them in a matrix form (rather than as a column list). The add-in also produces a heatmap of the correlations over time.

Now click the *Repeated Measures Cov/Corr Diagnostics* add-in using the button for this under your *Add-Ins* tab in the JMP menu bar. If you have several model reports open, you will be prompted to select the report on which you want to use this add-in. The add-in will produce additional output at the bottom of your report. Scroll down to find the correlation and covariance matrices (including a version of the covariance matrix that reports the standard errors for those variance and covariance estimates) and the correlation heatmap, shown in Figure 6.4.

If the default color gradient on the heatmap is not very informative, right-click over the red-to-blue color scale and select *Gradient*. Click in the *Color Theme* to select something *Sequential* (if you have only positive correlations, for example). Adjust the lightness range if desired. Change the *Range Type* to *Exact Data Range* to better distinguish

Figure 6.4: Unstructured Variances and Correlations

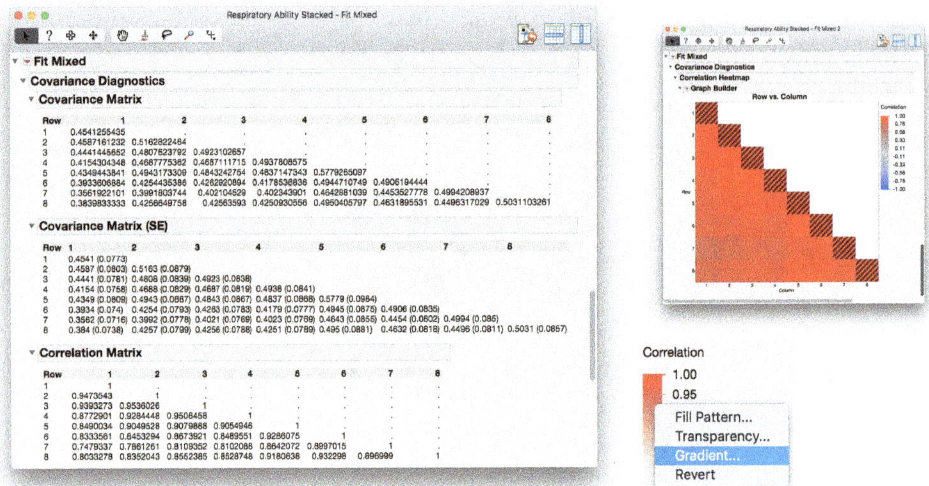

between high and low correlations for your specific range, as in Figure 6.5.

Figure 6.5: Color Themes for the Heatmap

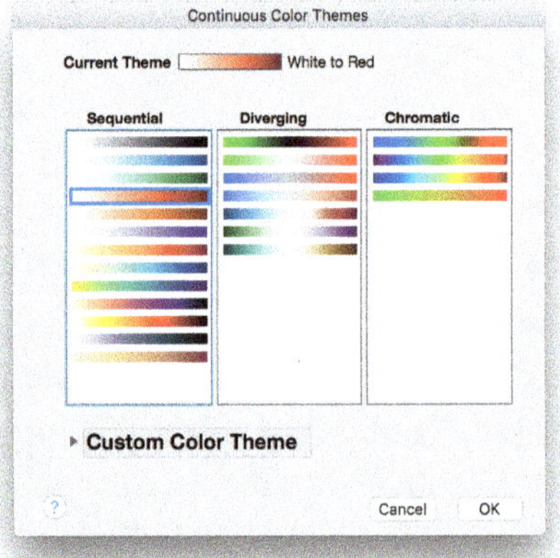

Now you can use the values in the diagonal of the *Unstructured Covariance Matrix* to determine whether the variances can be treated as equal or better treated as unequal.

Then use the *Correlation Heatmap*, shown in Figure 6.6, to determine whether there is a pattern as you move your gaze diagonally down and to the left.

Figure 6.6: Heatmap of Correlations

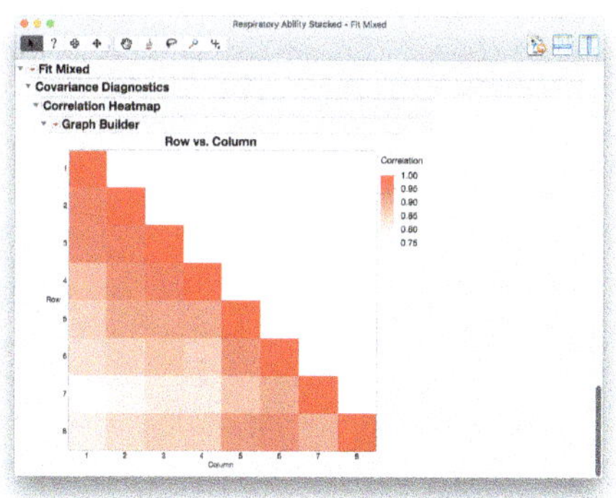

In this case, the variances for each time point (these are the values across the diagonal in the covariance matrix) range from 0.45 to 0.58. If we use the rule-of-thumb that standard deviations need to be at least a two-fold difference to require being treated as unequal, that would mean that these variances need to be at least four times different. By that standard, we can treat these variances as approximately equal across all eight time points. Likewise, look at the standard errors for these variances. If a 95% confidence interval for the first variance would be approximately $0.4541 \pm 2 * 0.0773$, that would be around $(0.30, 0.61)$. From this we deduce that any test of the difference between two of these variances will fail to reject the null hypothesis that the variances are equal. But, keep in mind that all of these heuristics are approximate and are actually trying to prove a null hypothesis (which we can't do!). It would be reasonable in this case to treat the variances as equal, but it might also be worth our time to try fitting unequal variances and see how that improves or worsens the fit (judged by fit statistics).

Looking at the correlation heatmap, it seems that the correlations are pretty high for small time lags, but they get systematically weaker as the time lag increases. From these observations we might consider the following options for the repeated structure:

- Compound Symmetry probably will not fit well, because it assumes no change in correlation over time lags.
- Toeplitz might fit well, although it ignores the decay relationship over the time lags.
- AR(1) (or, equivalently, Spatial Power) is a great choice, if that decay is in a power relationship.
- Antedependent is a possibility, because that is somewhat between Toeplitz and AR(1). But Antedependent will introduce a lot of new parameters, which might not be worth any small improvement in fit over a simpler model like AR(1). We can confirm this by looking at the fit statistics that penalize additional parameters (use AICc or BIC).

The resulting fit statistics of all four possible models are shown in Table 6.2. We are

Table 6.2: Model Fit Statistics for Respiratory Ability Model Contenders, all with Equal Variances

Repeated Structure	AICc	BIC
Compound Symmetry	364.8	475.5
AR(1)	297.0	407.7
Antedependent	282.3	417.8
Toeplitz	253.3	**388.8**
Unstructured	236.7	483.8

looking for a model that has low values on the fit statistics, relative to the other contenders. We can see that the Toeplitz is the best fitting model on the Bayesian Information Criterion (BIC), and it is approaching the Unstructured fit on the corrected Akaike's Information Criterion (AICc).

Extending the model: including the baseline measurements and quadratic effect for time

Using JMP Pro

Use *Analyze –> Fit Model* to define the model. Choose the *Mixed Model* Personality in order to see the tabs for *Construct Model Effects* that include *Random Effects* and *Repeated Structure*.

Enter **`FEV1`** as the *Y* variable. Enter **`basefev1`**, **`drug`**, **`hour as number`**, and **`drug*hour as number`** as *Fixed Effects*. To add the quadratic effect of **`hour as number`**, select this variable in the Columns List and choose *Macros > Polynomial to Degree*.

For the **Repeated Structure**, select *Unstructured* and enter `hour` as the *Repeated* and `patient(drug)` as the *Subject*.

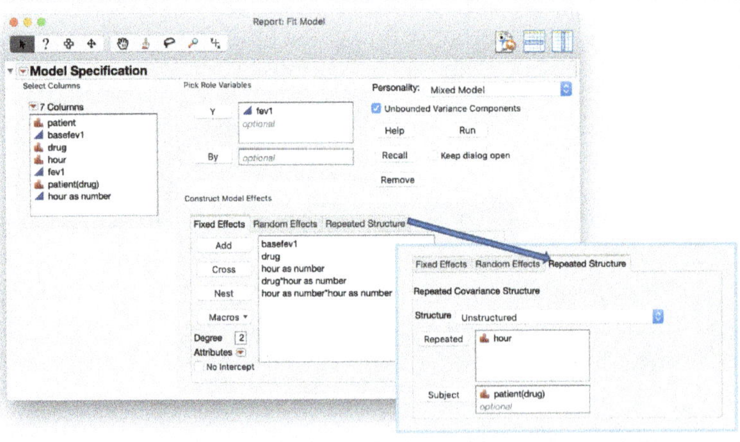

Let's now refit the model using the additional Baseline measurement as a covariate and adding a quadratic effect for time. Results of the variance and correlation output are shown in Figure 6.7.

Figure 6.7: Unstructured Variances and Correlations for Baseline+Polynomial Model

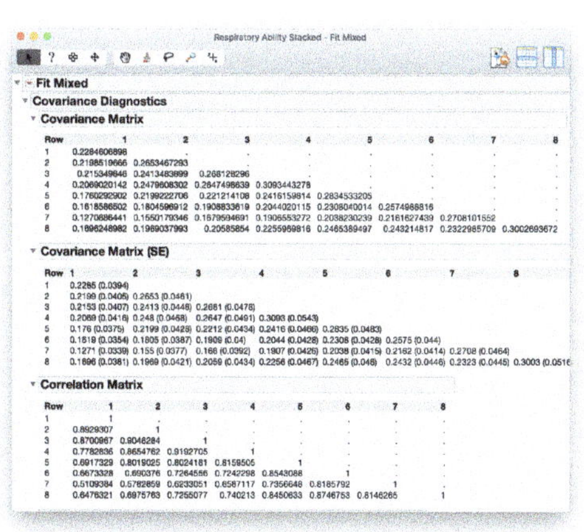

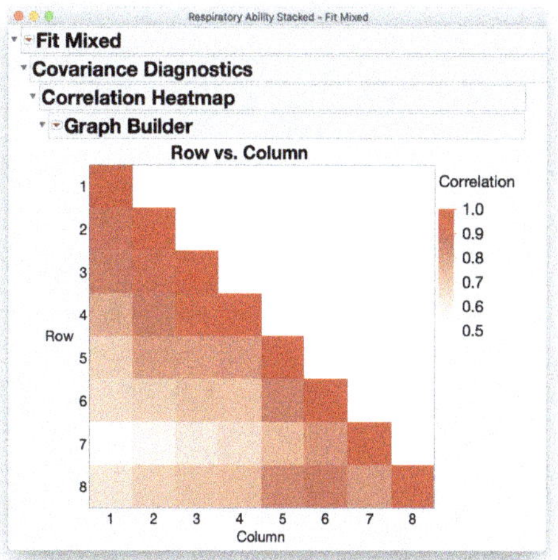

Again, the evidence suggests equal variances at all time points and a similar correlation across the time lags with a slight decay. Again, we fit the CS, AR(1), Antedependent, and Toeplitz structures, this time with our new treatment model.

The resulting fit statistics of the four models are shown in Table 6.3.

Again, the Toeplitz is the best fitting model on the Bayesian Information Criterion (BIC), and it is approaching the Unstructured fit on the corrected Akaike's Information Criterion (AICc).

Table 6.3: Model Fit Statistics for Respiratory Ability Model Contenders, All with Equal Variances

Repeated Structure	**AICc**	**BIC**
Compound Symmetry	300.4	343.6
AR(1)	227.3	270.5
Antedependent	212.3	281.0
Toeplitz	197.1	**265.8**
Unstructured	181.7	365.9

Interpretation of the Final Model

The model fit and test output based on the Toeplitz correlation structure is show in Figure 6.8.

Figure 6.8: Final Model Output

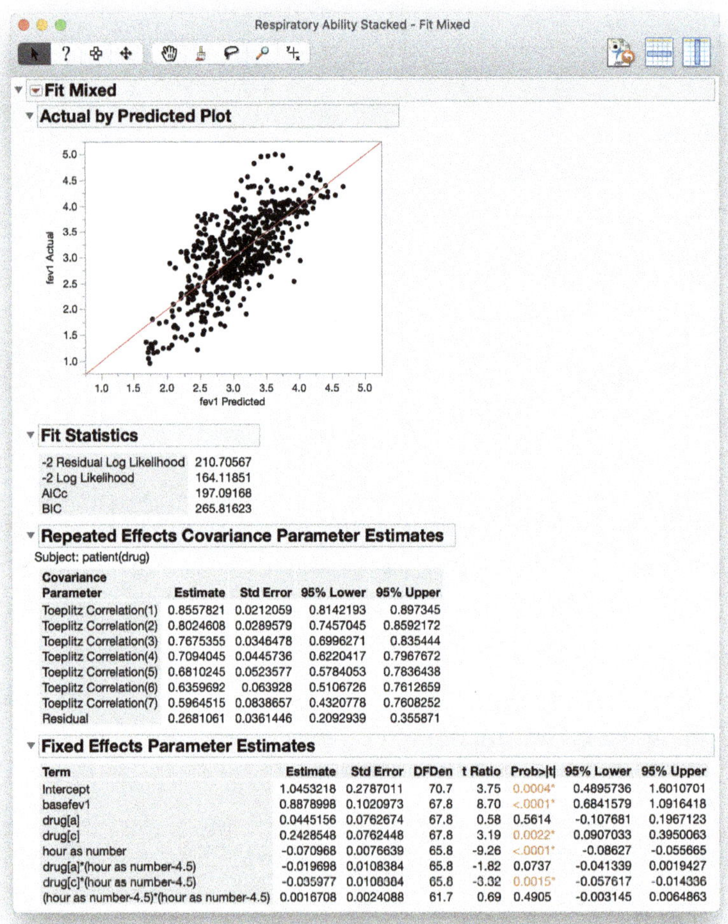

Fit Statistics

-2 Residual Log Likelihood	210.70567
-2 Log Likelihood	164.11851
AICc	197.09168
BIC	265.81623

Repeated Effects Covariance Parameter Estimates

Subject: patient(drug)

Covariance Parameter	Estimate	Std Error	95% Lower	95% Upper
Toeplitz Correlation(1)	0.8557821	0.0212059	0.8142193	0.897345
Toeplitz Correlation(2)	0.8024608	0.0289579	0.7457045	0.8592172
Toeplitz Correlation(3)	0.7675355	0.0346478	0.6996271	0.835444
Toeplitz Correlation(4)	0.7094045	0.0445736	0.6220417	0.7967672
Toeplitz Correlation(5)	0.6810245	0.0523577	0.5784053	0.7836438
Toeplitz Correlation(6)	0.6359692	0.063928	0.5106726	0.7612659
Toeplitz Correlation(7)	0.5964515	0.0838657	0.4320778	0.7608252
Residual	0.2681061	0.0361446	0.2092939	0.355871

Fixed Effects Parameter Estimates

Term	Estimate	Std Error	DFDen	t Ratio	Prob>|t|	95% Lower	95% Upper
Intercept	1.0453218	0.2787011	70.7	3.75	0.0004*	0.4895736	1.6010701
basefev1	0.8878998	0.1020973	67.8	8.70	<.0001*	0.6841579	1.0916418
drug[a]	0.0445156	0.0762674	67.8	0.58	0.5614	-0.107681	0.1967123
drug[c]	0.2428548	0.0762448	67.8	3.19	0.0022*	0.0907033	0.3950063
hour as number	-0.070968	0.0076639	65.8	-9.26	<.0001*	-0.08627	-0.055665
drug[a]*(hour as number-4.5)	-0.019698	0.0108384	65.8	-1.82	0.0737	-0.041339	0.0019427
drug[c]*(hour as number-4.5)	-0.035977	0.0108384	65.8	-3.32	0.0015*	-0.057617	-0.014336
(hour as number-4.5)*(hour as number-4.5)	0.0016708	0.0024088	61.7	0.69	0.4905	-0.003145	0.0064863

We see that, holding the Time and Baseline at their means (Time = 4.5 and Baseline = 2.6493), Drug C is statistically significantly different from the Placebo, but Drug A is not. We see that the trend in the response over time for the Placebo has an estimated slope of −0.071, which is statistically significantly different from zero ($p < 0.0001$), and that the time trend shrinks to −0.036 (still statistically significantly different from zero) for

Drug C and drops to a statistically non-significant trend estimated at −0.019 for Drug A. We see that the quadratic effect of time is not statistically significant.

From the Fit Mixed red triangle, choose *Multiple Comparisons* and select the *Tukey HSD* to see the comparison of the FEV1 means across the three Drug groups. This output is shown in Figure 6.9. The Tukey HSD Pairwise Comparisons show that Drugs A and C are not statistically different in their mean FEV1, averaged over all time points and over all Baseline measurements, but that Drugs A and C are both statistically significantly different from the Placebo for this comparison.

Figure 6.9: Final Model LSMeans

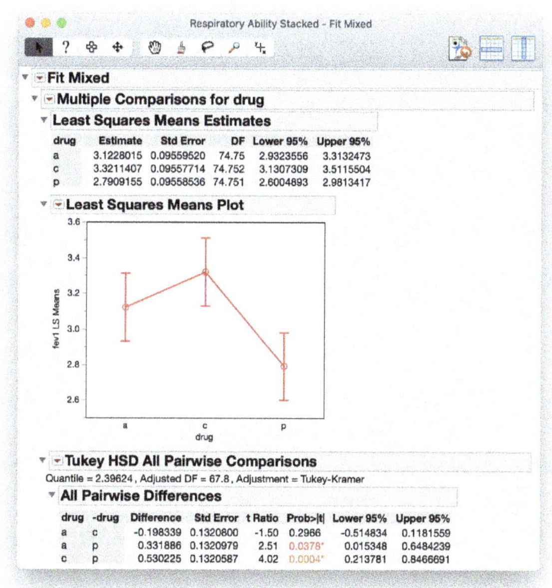

drug	Estimate	Std Error	DF	Lower 95%	Upper 95%
a	3.1228015	0.09559520	74.75	2.9323556	3.3132473
c	3.3211407	0.09557714	74.752	3.1307309	3.5115504
p	2.7909155	0.09558536	74.751	2.6004893	2.9813417

drug	-drug	Difference	Std Error	t Ratio	Prob>|t|	Lower 95%	Upper 95%
a	c	-0.198339	0.1320800	-1.50	0.2966	-0.514834	0.1181559
a	p	0.331886	0.1320979	2.51	0.0378*	0.015348	0.6484239
c	p	0.530225	0.1320587	4.02	0.0004*	0.213781	0.8466691

We are likely much more interested in fixing various levels of the Baseline and the Time and doing the comparisons within those conditions. We can explore this visually by selecting the *Marginal Model Inference > Profiler* from the Fit Mixed red triangle at the top of the model output. Several conditions are shown in Figure 6.10.

From the Profiler, we can see that Drugs A and C are pretty similar in mean FEV1, and slightly higher in FEV1 than Placebo, at the midpoints of Baseline and Time. However, at earlier times, Drugs A and C were even higher in comparison to Placebo than they were at the mid-time. By the end of the experiment, all three groups were really converging in FEV1 measurements. We can also explore this change in the Drug comparisons over time at different fixed values for the Baseline FEV1, as in the plot fixing Baseline

Figure 6.10: Final Model Marginal Means Profiler

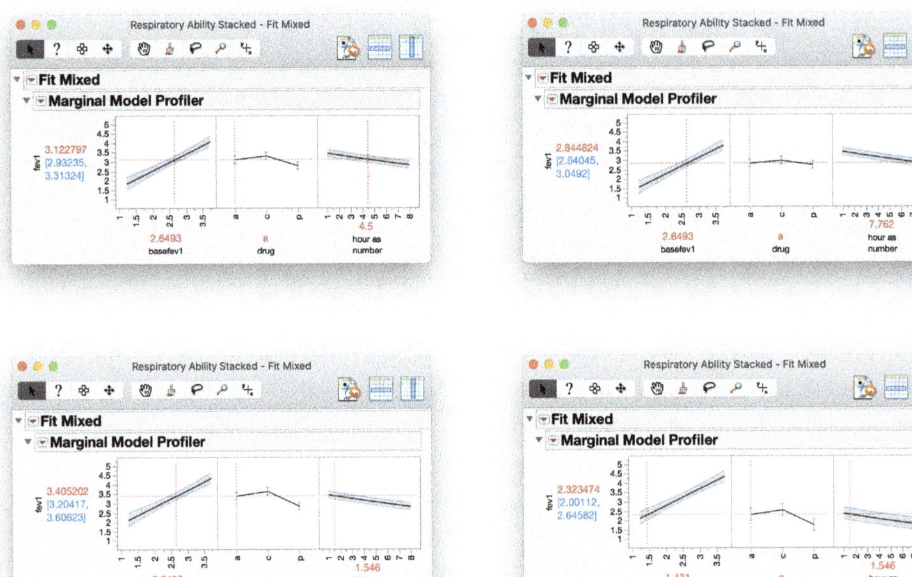

at a lower value.

We can also do statistical tests of comparisons for fixed values of the Baseline and Time. From the Fit Mixed red triangle, choose *Multiple Comparisons* and then choose *User-Defined Estimates*. The dialog box and some of the results are shown in Figure 6.11.

From the results in Figure 6.11, we can see that, for Baseline FEV1 of 2.5, Drug A at Hour 1 is statistically significantly different than at Hour 3, 5, or 8. For Baseline FEV1 of 2.5, Drug A and Drug C are not statistically different from each other at Hour 1 or at Hour 3 or at Hour 5.

Why Use a Baseline Measurement Rather Than Including It as Time=0?

We could have chosen to arrange these data with 9 rows per patient, with the first row corresponding to the baseline, or $Time = 0$, measurements. In this case, we could have followed the same procedures as described above, but excluding an additional effect for the baseline.

What is the benefit of treating the baseline measurements as a covariate, that is, as a separate non-random continuous term in the model rather than as a first time point in our repeated measures series?

Figure 6.11: Multiple Comparisons at Fixed Values

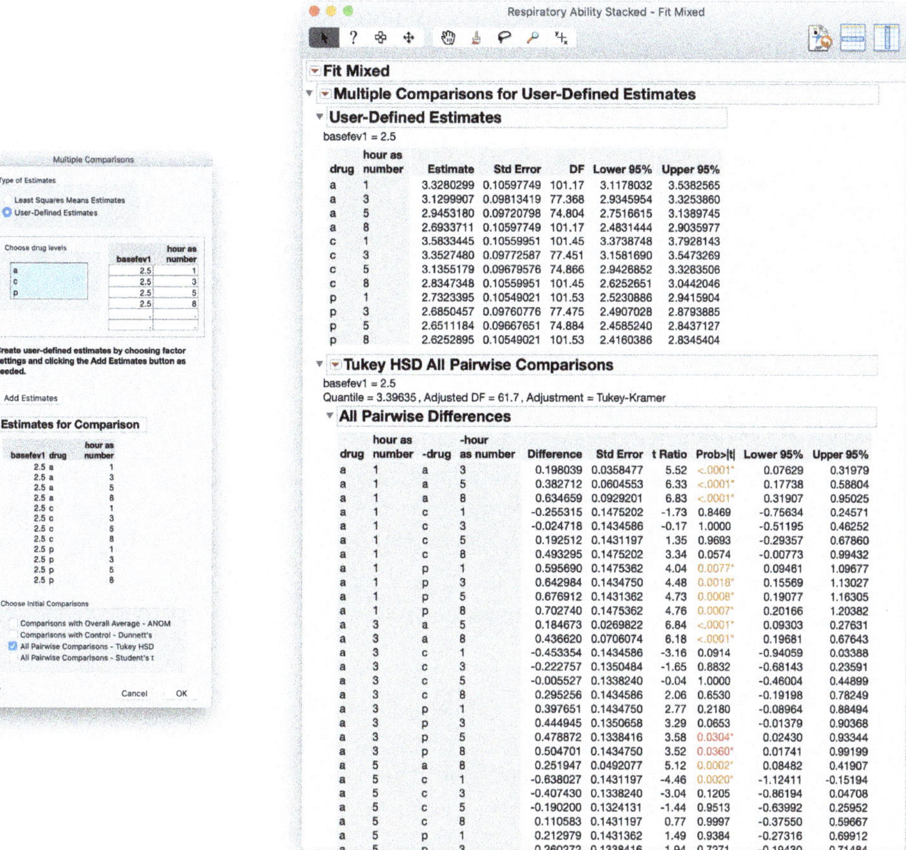

drug	hour as number	Estimate	Std Error	DF	Lower 95%	Upper 95%
a	1	3.3280299	0.10597749	101.17	3.1178032	3.5382565
a	3	3.1299907	0.09813419	77.368	2.9345954	3.3253860
a	5	2.9453180	0.09720798	74.804	2.7516615	3.1389745
a	8	2.6933711	0.10597749	101.17	2.4831444	2.9035977
c	1	3.5833445	0.10559951	101.45	3.3738748	3.7928143
c	3	3.3527480	0.09772587	77.451	3.1581690	3.5473269
c	5	3.1355179	0.09679576	74.866	2.9426852	3.3283506
c	8	2.8347348	0.10559951	101.45	2.6252651	3.0442046
p	1	2.7323395	0.10549021	101.53	2.5230886	2.9415904
p	3	2.6850457	0.09760776	77.475	2.4907028	2.8793885
p	5	2.6511184	0.09667651	74.884	2.4585240	2.8437127
p	8	2.6252895	0.10549021	101.53	2.4160386	2.8345404

drug	hour as number	-drug	-hour as number	Difference	Std Error	t Ratio	Prob>|t|	Lower 95%	Upper 95%
a	1	a	3	0.198039	0.0358477	5.52	<.0001*	0.07629	0.31979
a	1	a	5	0.382712	0.0604553	6.33	<.0001*	0.17738	0.58804
a	1	a	8	0.634659	0.0929201	6.83	<.0001*	0.31907	0.95025
a	1	c	1	-0.255315	0.1475202	-1.73	0.8469	-0.75634	0.24571
a	1	c	3	-0.024718	0.1434586	-0.17	1.0000	-0.51195	0.46252
a	1	c	5	0.192512	0.1431197	1.35	0.9693	-0.29357	0.67860
a	1	c	8	0.493295	0.1475202	3.34	0.0574	-0.00773	0.99432
a	1	p	1	0.595690	0.1475362	4.04	0.0077*	0.09461	1.09677
a	1	p	3	0.642984	0.1434750	4.48	0.0018*	0.15569	1.13027
a	1	p	5	0.676912	0.1431362	4.73	0.0008*	0.19077	1.16305
a	1	p	8	0.702740	0.1475362	4.76	0.0007*	0.20166	1.20382
a	3	a	5	0.184673	0.0269822	6.84	<.0001*	0.09303	0.27631
a	3	a	8	0.436620	0.0706074	6.18	<.0001*	0.19681	0.67643
a	3	c	1	-0.453354	0.1434586	-3.16	0.0914	-0.94059	0.03388
a	3	c	3	-0.222757	0.1350484	-1.65	0.8832	-0.68143	0.23591
a	3	c	5	-0.005527	0.1338240	-0.04	1.0000	-0.46004	0.44899
a	3	c	8	0.295256	0.1434586	2.06	0.6530	-0.19198	0.78249
a	3	p	1	0.397651	0.1434750	2.77	0.2180	-0.08964	0.88494
a	3	p	3	0.444945	0.1350658	3.29	0.0653	-0.01379	0.90368
a	3	p	5	0.478872	0.1338416	3.58	0.0304*	0.02430	0.93344
a	3	p	8	0.504701	0.1434750	3.52	0.0360*	0.01741	0.99199
a	5	a	8	0.251947	0.0492077	5.12	0.0002*	0.08482	0.41907
a	5	c	1	-0.638027	0.1431197	-4.46	0.0020*	-1.12411	-0.15194
a	5	c	3	-0.407430	0.1338240	-3.04	0.1205	-0.86194	0.04708
a	5	c	5	-0.190200	0.1324131	-1.44	0.9513	-0.63992	0.25952
a	5	c	8	0.110583	0.1431197	0.77	0.9997	-0.37550	0.59667
a	5	p	1	0.212979	0.1431362	1.49	0.9384	-0.27316	0.69912
a	5	p	3	0.260272	0.1338416	1.94	0.7271	-0.19430	0.71484

In this case, the *Baseline* = 0 choice is because we don't actually know the precise day of the baseline measurement. The baseline might not have happened at *Time* = 0; it was taken during a range of times before the randomization to the drug arm. In this case, we don't want to force the time spacing to be untrue for this measurement, so instead we can use it as a covariate.

6.5 When Might We Choose Other Models?

Experiments involving repeatedly measuring subjects over time or varying experimental conditions will not always be best modeled using this repeated-measure covariance-adjusted mixed model framework. Sometimes the research question, other assumptions about the model, or another aspect of the experiment design will be better fit in another way.

As we discussed above, using the Unstructured option in the repeated structure is equivalent to using MANOVA, and using the CS option in the repeated structure is equivalent to using the Subject as a random effect in a standard mixed model. There are other modeling options, too. When might another model be appropriate?

Nonlinear Curve-Fitting

When you want to fit a special non-linear curve to the trend over time, such as a 4-parameter logistic model in pharmacokinetics, you can use the *Analyze > Specialized Modeling > Nonlinear* platform in JMP. However, there is no built-in Nonlinear Mixed Model capability in JMP, so random effects will not be allowed. In this case, you can use the connection between JMP and SAS to pass data and write and run a SAS PROC NLMIXED program. You can learn more about using JMP and SAS together at SAS Institute Inc. (2019b).

Random Coefficients

Combining standard curve-fitting (with polynomials and powers, rather than using specialized nonlinear formulas) with the random effect for each subject is possible using a Random Coefficients model. This is covered in Chapter 5. Combining random coefficients with a correlation structure is often problematic due to confounding and identifiability issues. See Stroup (2012) for more information about this topic.

Time Series

The time element, when measuring something repeatedly, can also be used to capture more specific linear and cyclical trends over time or to forecast predictions into the future. Time series methods, available from the *Analyze > Specialized Modeling > Time Series* and *Time Series Forecast* platforms, will be useful in this case. However, random effects are not allowed in Time Series models in JMP. This is another opportunity to use the connection from JMP to SAS (or R or Python) to access additional models that may be available in these environments.

Multiple-Conparisons Adjustments

In the introduction to this chapter, and again in Section 6.2, we mentioned that "repeated measures" can also describe distinctly different measurements taken on the same subject. This is extremely common in experimenting: the researcher sets up the experiment design conditions and applies the treatments and then records several responses for each subject.

As with repeated measures of the same response over time, the modeling concern is about addressing the correlation between measurement on the same subject. MANOVA is one approach to allow that correlation structure to be modeled, but it imposes severe restrictions of *sphericity*, which are often not met, on the joint distribution of all of the responses. The restriction of sphericity means that the variances of the differences, within subjects, between all possible pairs of the various responses are equal. This restriction is unreasonable when, for example, some measurements are continuous and others are indicators of presence/absence of a trait.

An alternative is to build the models separately for each response, collect the p-values for the planned comparisons, and apply an FDR multiple comparisons adjustment to control the overall experiment-wide error rate. For multiple responses taken on the same subjects, with the same planned model, this can be done in the Fit Model platform in JMP by adding all responses into the "Y, Response" box, running this with the Standard Least Squares Personality (and do not select "Fit Separately"), and selecting the "FDR" box below the table of p-values in the Effects Summary at the top, as shown in Figure 6.12.

Figure 6.12: FDR adjustment for multiple responses

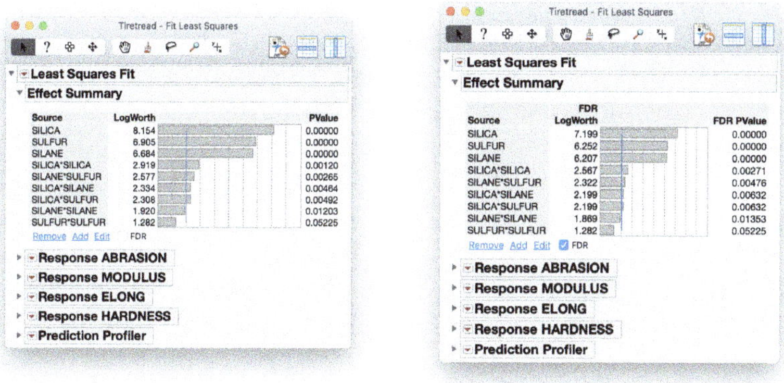

In the case that some responses were collected at a block-level (rather than at the subject-level) or that the planned model requires a different *personality* option in JMP (for example, *Mixed Model*), there is a JMP add-in that collects the p-values from all open model reports, creates a table of these, and applies the FDR adjustment (Parris and Hummel, 2021a). You can download that JMP add-in at `community.jmp.com`.

When Do We Choose the Methods in This Chapter?

The methods of this chapter are most appropriate when the primary goal is to find differences in the response between time points and among interactions of time with various treatment conditions. The alternate methods may be a better fit when the primary goal is instead to explore many unrelated measurements on the same subjects, or to fit a curve to each subject's trajectory, or to find the average trend curve over time, or to forecast into the future.

6.6 Exercises

1. In the **Cholesterol Stacked.jmp** data table, there are five subjects in four treatment groups, with measurements taken in the morning and afternoon, once a month, for three months. The goal is to fit a model for the response, Y, based on the Treatment, Month, and AM/PM as full factorial effects.
   (a) Follow the steps described in this chapter to fit first the Unstructured Covariances, plan out reasonable candidates for other structures to try, and fit these. Compare fit statistics to choose a best model.
   (b) Next, fit a model ignoring the covariances inherent in the repeated measures. Instead, include a random effect for Patient. (Note that Patient 1 in Treatment A is not the same person as Patient 1 in Treatment B. This requires us to either rename the Patients with unique identifiers or to use Patient[Treatment] (read as "Patient nested in Treatment") when referring to the Patient effect in the model.) This model was a common historical model for repeated measures, accounting for the similarity of measures within Patient by making that a random block effect, but ignoring anything else about the repeated structure.
   (c) Compare the Fixed Effects results for Treatment, Month, AM/PM, and their interactions, on this best model compared to the model that does not account for the repeated measure correlations and covariances.
2. In this exercise, you will compare the results of analyses using MANOVA, Unstructured Covariance, and FDR adjusted comparisons on **Tiretread.jmp** and **Tiretread Stacked.jmp**.

   These data are from Derringer and Suich (1980). Four characteristics (Abrasion, Modulus, Elongation, and Hardness) were measured on each of twenty tires. The tires consist of various amounts of Silica, Silane, and Sulfur. The experiment was arranged as a Response Surface design in order to explore the main treatment effects, the two-way treatment interactions, and the treatment quadratic effects.
   (a) Open **Tiretread.jmp**. Run the Response Surface model for all four responses. (You can use the saved script "RSM for 4 Responses" to do this.) In the combined Effect Summary at the top of the output, which effects are statistically significant at the $\alpha = 0.05$ level? Notice the FDR option below this table. You can select this to do an FDR adjustment for multiple comparisons to this table of tests. Does anything change dramatically in p-value?

   This combined Effect Summary takes the minimum p-value for each effect across all four individual responses, so this tells us that any effect that is statistically significant in the combined Effect Summary will be statistically significant in at least one of the comparisons below.

   Now look at the analysis for just the Abrasion response, immediately below the combined Effect Summary table. Which responses are statistically significant at the $\alpha = 0.05$ level for just the Abrasion response? Look at the results for the other three responses and identify the statistically significant effects for these, also.

You can also use the FDR add-in referenced in Section 6.5 to do an FDR adjustment for multiple comparisons. However, this is not needed for a model with only continuous responses.

(b) Open **`Tiretread Stacked.jmp`**. Notice that this arrangement of the data has stacked each of the four responses on top of each other, creating four rows per tire. The new Characteristic column now describes which response was measured. Run the saved script "Fit Model – Unstructured" and reopen the Model Dialog box that generated this model (by clicking the red triangle for the Fit Mixed output and selecting *Model Dialog*). Notice that the new model effects list includes, in addition to the original Response Surface model, Characteristic as a main effect, as well as Characteristic interacting with all of the Response Surface model effects. This allows us to compare the significance of the various Response Surface model effects across the different Characteristics.

Hover your mouse over the heading row of the table of Fixed Effects Parameter Estimates. Right-click and select *Sort by Column*. Choose *Prob>|t|* and click *OK*. The bottom of the table shows the statistically significant effects. We can see that HARDNESS and ABRASION have statistically significantly different means compared to the MODULUS average response. This should makes sense when you think about the scale of the measurements: Modulus ranges from 490 to 2294, which is a bit similar to Elongation, which ranges from 240 to 640. However, Hardness ranges from 62.5 to 78, and Abrasion goes ranges form 96 to 198. Note that JMP chooses the last group, alphanumerically, as the group to which comparisons are made. Since Modulus is the last in an alphabetical ordering of the Characteristic names, it is the baseline group. You can change this baseline group, if desired by, adding Value Ordering in the column properties for Characteristic in the data table.

Look at the other statistically significant effects. Silane, Silica, and the quadratic of Sulfur, are all statistically significant for Modulus. Which effects have an additional statistically significant effect for only one or some of the other characteristics?

(c) Explore the results a bit more. For the Stacked analysis, look at the *Profiler* and *Surface Profiler* under *Marginal Model Inference* from the top red triangle. Fix the value for one Characteristic and explore the effect of the Silane, Sulfur, and Silica for that response. On the Surface Profiler, select Characteristic as the Y. Notice that the Characteristics are called 0, 1, 2, and 3, corresponding to Abrasion, Elongation, Hardness, and Modulus. Look at the Silica on the X axis. How is the effect of Silica different on the different Characteristics?

(d) How do the results and interpretation differ for the four-univariate-responses model with FDR-adjusted p-values versus the single model that adjusts for the estimated covariance structure?

Chapter 7

Spatial Models

In the previous chapter, we discussed fitting covariance structures when observations were taken multiple times on the experimental units, either at several times or with multiple responses. Just as observations can be correlated across time, so can observations that are near each other in space. Often this similarity in response is accounted for through blocking criteria, but what can we do if there are no obvious blocking factors to use or if blocking did not "do the job"? Spatial covariance structures can help. Because spatial covariance structures are only available in the Mixed Model personality of Fit Model, spatial models are available in JMP Pro only.

7.1 Motivating Examples

Hazardous Waste — The investigator for an environmental study wants to evaluate water drainage at a potential hazardous waste disposal site. It is believed that water movement is affected by the thickness of a layer of salt, a geological feature of the area. Samples taken closer together are more alike than those farther apart.

Wheat Trial — A researcher wants to compare fifty-six varieties of wheat. A blocked design is frequently used for this type of experiment, so four complete blocks were planted in a field. Initial statistical results from the RCB trial do not "make sense" to the subject-matter expert. A different model may produce sensible conclusions.

7.2 Conceptual Background

Much of the terminology used with spatial models comes from geostatistics, specifically in mining applications. In depth discussions of this history include Journel and Huijbregts (1978), Isaaks and Srivastava (1989), Cressie (1991), and Schabenberger and Gotway (2005). We will explain the key terms here and how they relate to the JMP output in the examples.

As with repeated measures in *time* discussed in Chapter 6, the random error in the model includes the correlation between nearby points in space. This covariance, $Cov(e_i, e_j) = \sigma^2(f(d_{ij}))$, is a function, $f()$, of the distance, d_{ij}, between two locations. There are several options for the covariance function, f, and deciding which to use is a part of the model fitting process.

There are two further complications to selecting the spatial covariance structure for the model. The first is related to the covariance function $f(d_{ij})$. If the function does not depend on the direction of the distance, the structure is *isotropic*. For example, if there are two observations one unit apart in distance east and west, then whether the distance is calculated by "moving east" or "moving west" does not affect the covariance value. If there is dependence upon the direction, then the structure is *anisotropic*. This often happens if there are prevailing winds or water flows. The spatial structures available in JMP Pro include both isotropic and anisotropic versions.

The second complication is referred to as the *nugget*. This nugget accounts for abrupt changes over relatively small distances, and the term comes from geostatistics. Imagine that you are digging for gold. You dig for a while with no evidence of gold and then the next shovel hits the gold deposit. In that short distance, you went from no gold to gold and found the nugget that you were seeking. How can we know whether there is this nugget effect? A visualization called a *variogram* will help.

A variogram (sometimes referred to as a semivariogram) plots the semivariance (the variance divided by two) of observations at various distances. When the observations are close together, you expect the points to be similar, and so the variance will be close to zero. When the observations increase in distance, the variability will increase. Typically, this variance plateaus at some distance. The value of this plateau is referred to as the *sill*. The distance that it takes to reach the sill is called the *range*. Observations further apart than the range are effectively uncorrelated. Figure 7.1 shows an empirical variogram of one of our examples.

In data where there is no nugget, the variogram will be close to zero near distance zero. In this example, it appears that the semivariance at that small distance is around 20, not 0, indicating there is likely a nugget effect. The variance increases as distance increases to about 20, then levels off. Empirically, the range is about 20, and the sill is around 75. The last two points on the variogram seem to indicate more variability at the greater distances, but there are not very many observations at that large distance apart, so the variance is inflated.

Finally, in some cases researchers are interested in estimating the spatial relationship itself. In other cases the research of interest is in estimating mean or regression effects accurately adjusting for the spatial effect. JMP Pro is better suited to the adjustment tasks due to scaling in some of the spatial parameter estimates.

Figure 7.1: Empirical Variogram

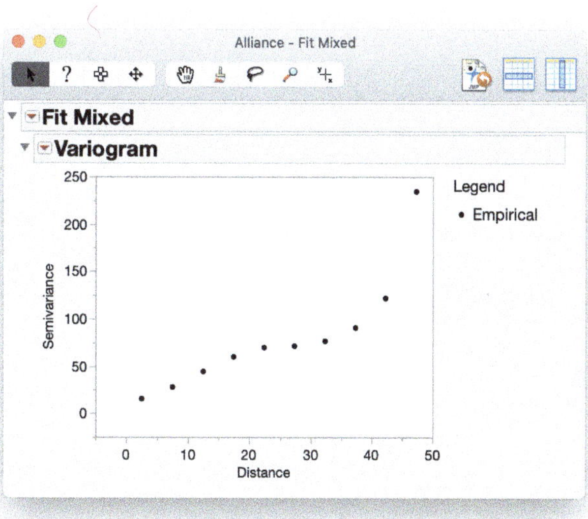

7.3 Hazardous Waste Example

In this environmental study, the investigator wants to evaluate water drainage characteristics at a potential hazardous waste disposal site. It is believed that water movement is affected by the thickness of a layer of salt, a geological feature of the area. Thirty samples were taken at various locations. The locations are identified by their north-south and east-west coordinates (`northing` and `easting`, respectively). The water movement was measured in log-transmissivity, `logt`. The researcher believes that there is a linear relationship between the salt thickness, `salt`, and `logt`. It is also assumed that samples taken closer together will be more correlated with each other than samples taken farther apart. These data first appeared in Littell et al. (1996) and then again in Littell et al. (2006) with thanks to Carol Gotway Crawford.

Experiment Design		**Skeleton ANOVA**	
Source	***df***	**Source**	***df***
		Salt	1
Sample	29	Sample|Salt	29-1 = 28
Total	29	Total	29

Model for the Hazardous Waste Data

$$logt = \beta_0 + \beta_1 salt + e$$

logt is the continuous response variable.
salt is the salt thickness.
β_0 is the intercept.
β_1 is the regression effect.
$\mathbf{e}$ is the error and $\mathbf{e} \sim N(\mathbf{0},\mathbf{R})$ where $\mathbf{R}$ is a spatial covariance structure.

Spatial Model Instructions for Hazardous Waste

With the `Hazardous Waste.jmp` data table open, go to *Analyze > Fit Model*. Enter `logt` as the *Y* variable. Change the default personality from *Standard Least Squares* to *Mixed Model*.
Select `salt` in the *Select Columns* box then click *Add* to add fixed effect of salt to the *Construct Model Effects* box.
In this example, there is nothing to enter on the *Random Effects* tab.
In the *Repeated Structure* tab of the *Construct Model Effects* box choose *Spatial* from the Structure drop-down box. Choose *Exponential* from the Type drop-down box. Enter `easting` and `northing` in the *Repeated* box. There is no subject in this model as the pair of `easting` and `northing` coordinates uniquely identify an observation.
Because choosing a spatial structure can be an iterative process, check the box for *Keep dialog open* to make returning to the model dialog box for further fits easier.
Click *Run* to fit the model.

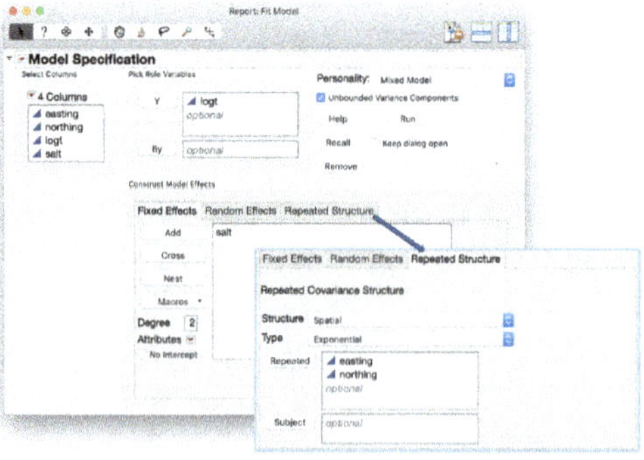

Results and Interpretation

Although the JMP report produced contains all of the usual fit information, we want to first assess the spatial fit from the variogram to determine whether the exponential structure seems reasonable and whether we should consider a nugget effect. Figure 7.2 shows the variogram portion of the report.

Figure 7.2: Variogram of Hazardous Waste Data with Exponential Fit

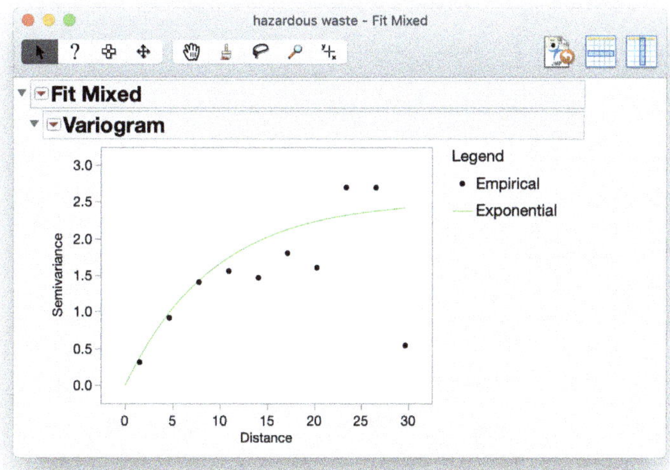

Other spatial structures can be added to the variogram to visually assess which might provide a better fit. Using the Variogram red triangle menu, add *Spatial with Nugget > Exponential* and *Spatial with Nugget > Spherical* to the Variogram. The two structures that include the nugget effect appear to fit the empirical variogram points a bit better. The Spherical structure clearly finds a sill at around semivariance of 1.5 with the range of about 10. Figure 7.3 includes these fits.

Figure 7.3: Variogram of Hazardous Waste Data with Multiple Fits

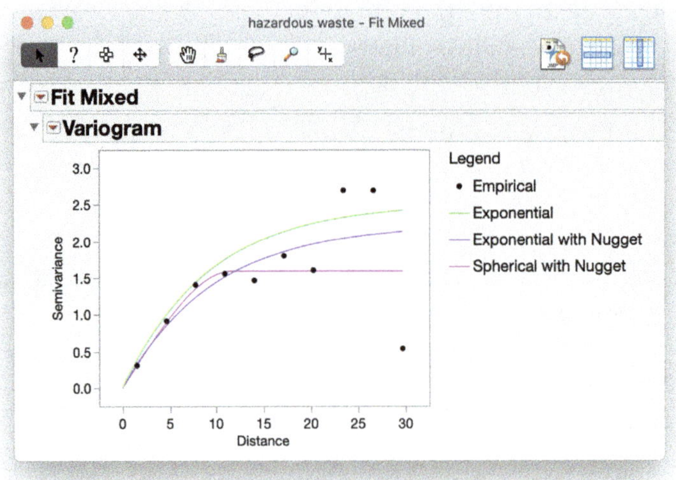

Returning to the Model dialog box, on the Repeated Structure tab change the Structure to *Spatial with Nugget* and the Type to *Spherical* because it looked optimal. Then click *Run* to fit this model. Although the variogram gave us visual indication that the spherical structure with a nugget might be more appropriate, when we compare the fit statistics between the two, the data tell a different story. The exponential structure has a lower AICc and BIC indicating a better fit. Figure 7.4 shows the results. The improvement in the -2 Log Likelihood of the nugget model is not enough to offset the penalty of adding another parameter to the estimation.

Figure 7.4: Fit Statistics for the Exponential Model versus the Spherical with Nugget Model

Fit Statistics

-2 Residual Log Likelihood	79.486199
-2 Log Likelihood	72.352991
AICc	81.952991
BIC	85.95778

Fit Statistics

-2 Residual Log Likelihood	79.896366
-2 Log Likelihood	71.535506
AICc	84.035506
BIC	88.541493

Table 7.1 includes fit statistics for a few other spatial structures of interest. In this case the exponential structure with no nugget is the optimal choice. Notice how the two structures with a nugget have essentially the same fit statistics. This is often the case. Once the spatial covariance is accounted for, the particular structure of that accounting does not necessarily matter.

Table 7.1: Model Fit Statistics for Hazardous Waste Model Contenders

Repeated Structure	**AICc**	**BIC**
Exponential	81.953	85.958
Spherical	83.237	87.242
Exponential with Nugget	84.572	89.078
Spherical with Nugget	84.056	88.541

Returning to the original exponential structure fit, we have the results to answer the researcher's question about the relationship between the log-transmissivity of water and salt depth. Figure 7.5 shows the full results of this model. The actual by predicted plot does not show a great fit, but the linear fixed effects model is the known appropriate model. The "Repeated Effects Covariance Parameter Estimates" table shows the estimated range and sill parameters of the exponential spatial structure. The range parameter, Spatial Exponential, is 9.27. The sill parameter, Residual, is 2.53. Looking back at the variogram in Figure 7.1, this value of the sill seems reasonable. The practical range for these data is about 28, (3×9.27), which also seems reasonable from the variogram.

Finally, the "Fixed Effects Parameter Estimates" contains the regression parameters of interest. The final model for the researcher is $logt = -5.029 - 0.0176 * salt$. With a p-value of 0.0056 there is strong evidence this model is better than an intercept only model.

Figure 7.5: Hazardous Waste Final Model Fit

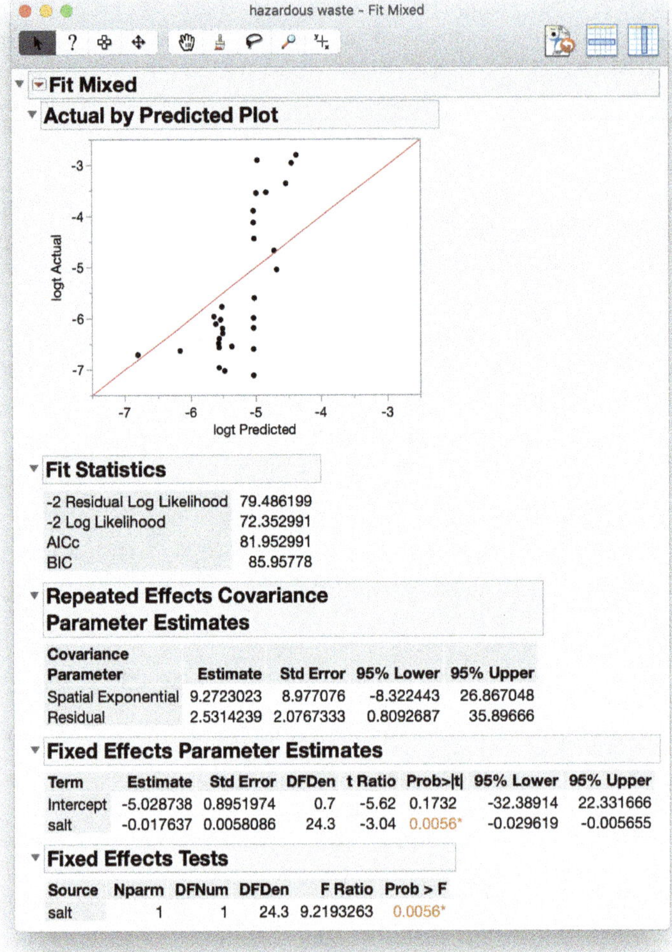

7.4 Alliance Wheat Trial

An agronomy researcher wanted to compare yield of 56 varieties of wheat. The initial experiment design was a randomized complete block design with four blocks. This experiment design led to problems in the analysis that were documented in Stroup et al. (1994). We will work through the analysis process used by the researchers to see how using spatial methods resolve the problems in the initial experiment design of this research. These data are in the data table `Alliance.jmp`.

The initial skeleton ANOVA for this experiment looks like this.

Experiment Design		**Skeleton ANOVA**	
Source	***df***	**Source**	***df***
Block	4-1=3	Block	3
		Variety	56-1=55
Plot(Block)	(56-1)*4=220	Residual	220-55=165
Total	224-1=223	Total	223

Initial RCB Model for the Alliance Data

$$y_{ij} = \mu + r_i + \tau_j + e_{ij}$$

y_{ij} is the yield on the ij^{th} observation.
μ is the overall intercept.
r_i is the effect of the ith block. $r_i \sim N(0, \sigma_b^2)$
τ_j is the effect of the j_{th} variety, `Entry`.
e_{ij} is the error of the ij_{th} observation and $e_{ij} \sim N(0, \sigma^2)$

This is the standard RCB model. However, the point of blocking is to have homogeneity of experimental units within the block. With block sizes of 56, there may not be homogeneity in the four blocks. This is potential problem number one. We continue with fitting the RCB model.

RCB Instructions for Alliance Wheat Trial

With the `Alliance.jmp` data table open, go to *Analyze* > *Fit Model*. Enter `Yield` as the *Y* variable. Change the default personality from *Standard Least Squares* to *Mixed Model*.
Select `Entry` in the *Select Columns* box then click *Add* to add fixed effect of variety to the *Construct Model Effects* box.
Enter `Rep` in the *Random Effects* tab of the *Construct Model Effects* box.
As this is the first of multiple models, check the box for *Keep dialog open* to make returning to the model dialog box for further fits easier.
Click *Run* to fit the model.

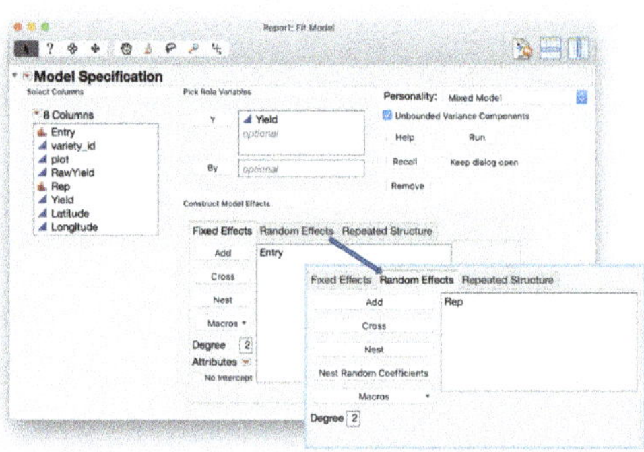

After the model is fit, add the least squares means using the *Multiple Comparisons* option in the red triangle menu. Once the Least Squares Means Estimates table is added, right-click on the table and choose *Sort by Column...*. Then choose *Estimate* as the column to sort by and click *OK*.

Results and Interpretation

Figure 7.6 shows selected results from the RCB model fit and LSMeans. At this point the subject matter expert knew something was not correct about this model. First, the overall Fixed Effects Test shows no statistical evidence of a difference between varieties when clearly in the field some varieties were superior to others. This on its own is not necessarily proof that the model is incorrect. Perhaps with more blocks "statistical significance" could be obtained. The estimates of the varieties' yields complete the picture.

The variety BUCKSKIN is known to be a superior performing variety, and in the field, it was obvious that it was doing better than the others. However, in our sorted list BUCKSKIN is in the middle of the 56 varieties, not at the top. The researcher was certain this could not be true given historical knowledge and field observation. Another model had to be found.

Figure 7.6: Alliance Wheat Trial RCB Model Fit

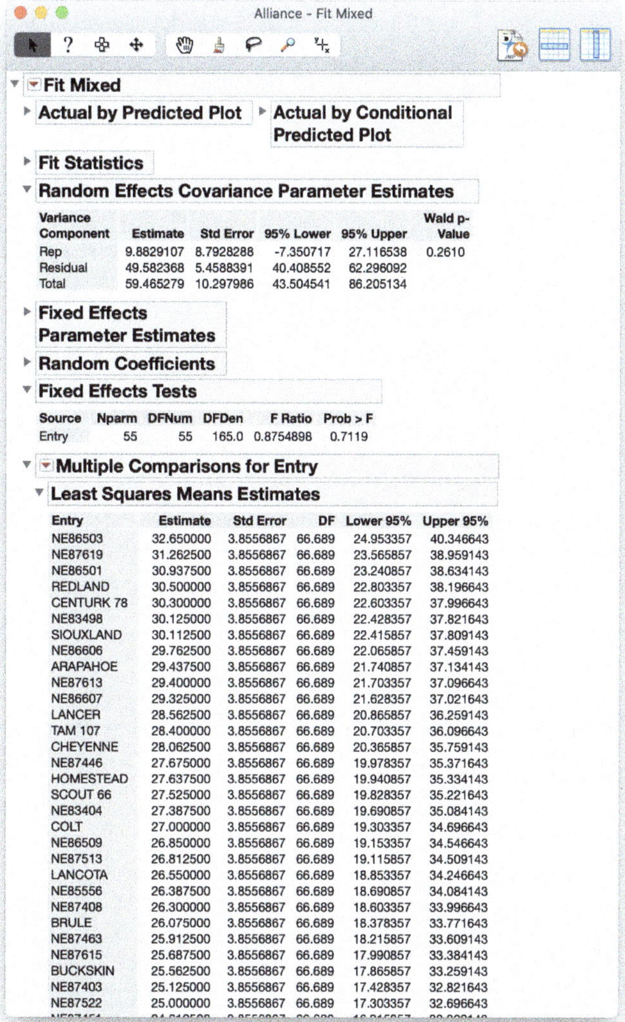

Alliance - Fit Mixed

Fit Mixed

Actual by Predicted Plot

Actual by Conditional Predicted Plot

Fit Statistics

Random Effects Covariance Parameter Estimates

Variance Component	Estimate	Std Error	95% Lower	95% Upper	Wald p-Value
Rep	9.8829107	8.7928288	-7.350717	27.116538	0.2610
Residual	49.582368	5.4588391	40.408552	62.296092	
Total	59.465279	10.297986	43.504541	86.205134	

Fixed Effects Parameter Estimates

Random Coefficients

Fixed Effects Tests

Source	Nparm	DFNum	DFDen	F Ratio	Prob > F
Entry	55	55	165.0	0.8754898	0.7119

Multiple Comparisons for Entry

Least Squares Means Estimates

Entry	Estimate	Std Error	DF	Lower 95%	Upper 95%
NE86503	32.650000	3.8556867	66.689	24.953357	40.346643
NE87619	31.262500	3.8556867	66.689	23.565857	38.959143
NE86501	30.937500	3.8556867	66.689	23.240857	38.634143
REDLAND	30.500000	3.8556867	66.689	22.803357	38.196643
CENTURK 78	30.300000	3.8556867	66.689	22.603357	37.996643
NE83498	30.125000	3.8556867	66.689	22.428357	37.821643
SIOUXLAND	30.112500	3.8556867	66.689	22.415857	37.809143
NE86606	29.762500	3.8556867	66.689	22.065857	37.459143
ARAPAHOE	29.437500	3.8556867	66.689	21.740857	37.134143
NE87613	29.400000	3.8556867	66.689	21.703357	37.096643
NE86607	29.325000	3.8556867	66.689	21.628357	37.021643
LANCER	28.562500	3.8556867	66.689	20.865857	36.259143
TAM 107	28.400000	3.8556867	66.689	20.703357	36.096643
CHEYENNE	28.062500	3.8556867	66.689	20.365857	35.759143
NE87446	27.675000	3.8556867	66.689	19.978357	35.371643
HOMESTEAD	27.637500	3.8556867	66.689	19.940857	35.334143
SCOUT 66	27.525000	3.8556867	66.689	19.828357	35.221643
NE83404	27.387500	3.8556867	66.689	19.690857	35.084143
COLT	27.000000	3.8556867	66.689	19.303357	34.696643
NE86509	26.850000	3.8556867	66.689	19.153357	34.546643
NE87513	26.812500	3.8556867	66.689	19.115857	34.509143
LANCOTA	26.550000	3.8556867	66.689	18.853357	34.246643
NE85556	26.387500	3.8556867	66.689	18.690857	34.084143
NE87408	26.300000	3.8556867	66.689	18.603357	33.996643
BRULE	26.075000	3.8556867	66.689	18.378357	33.771643
NE87463	25.912500	3.8556867	66.689	18.215857	33.609143
NE87615	25.687500	3.8556867	66.689	17.990857	33.384143
BUCKSKIN	25.562500	3.8556867	66.689	17.865857	33.259143
NE87403	25.125000	3.8556867	66.689	17.428357	32.821643
NE87522	25.000000	3.8556867	66.689	17.303357	32.696643

Plotting the data using the geographic coordinates of the field locations provides some insight.

Graph Builder Instructions for Alliance Wheat Trial

With the `Alliance.jmp` data table open, go to *Graph* > *Graph Builder*. Drag `Longitude` to the *X* drop zone. Drag `Latitude` to the *Y* drop zone. Drag `Yield` to the *Color* drop zone.
Click the *Heatmap* icon to change the graph type to a heat map.
Click *Done* to finish the graph.

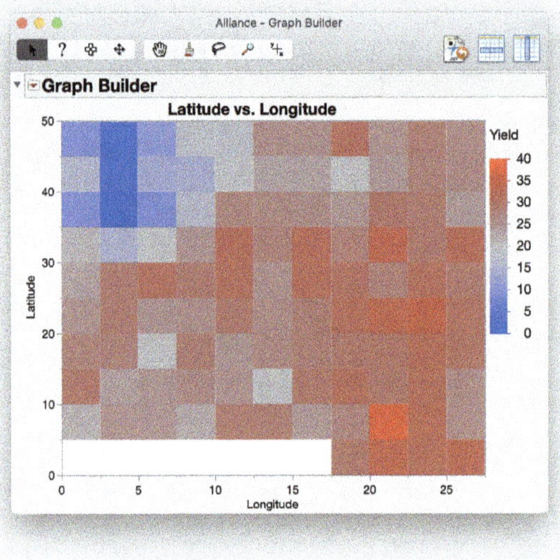

In the upper left of the heatmap, the yields are much less than toward the lower right. This is clear indication of a spatial process of some kind. Coincidentally, the randomization of the varieties within the blocks placed two replications of the BUCKSKIN variety in the lower yield zone, and the other two replications were also in the lower latitudes. The variability indicated by the heatmap is plausible biologically due to irregular fertility gradients from the spatial process.

Two models could be tried to include to the spatial variability. Either the original RCB model with the errors, e_{ij} following a spatial covariance structure, or omitting the block term and using a spatial model only. We will start with the RCB model and add the spatial spherical structure to model the spatial correlation present in the data.

Spatial Model Instructions for Alliance Wheat Trial

If you left the Model dialog box open, skip to entering the `Latitude` and `Longitude` in the *Repeated Structure* tab. Otherwise, you can click *Recall* to recall the RCB model or follow the instructions below.

With the `Alliance.jmp` data table open, go to *Analyze > Fit Model*. Enter `Yield` as the *Y* variable. Change the default personality from *Standard Least Squares* to *Mixed Model*.

Select `Entry` in the *Select Columns* box then click *Add* to add fixed effect of variety to the *Construct Model Effects* box.

Enter `Rep` in the *Random Effects* tab of the *Construct Model Effects* box.

In the *Repeated Structure* tab of the *Construct Model Effects* box choose *Spatial with Nugget* from the Structure drop-down box. Choose *Spherical* from the Type drop-down box. Enter `Latitude` and `Longitude` in the *Repeated* box.

Click *Run* to fit the model.

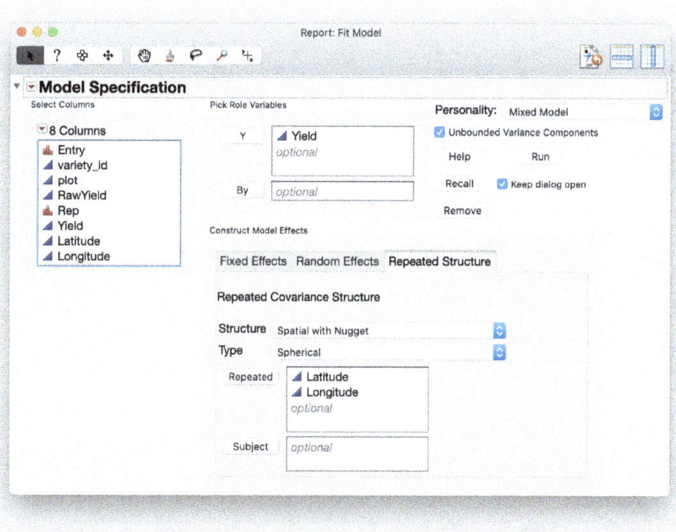

Results and Interpretation

The JMP report in Figure 7.7 presents a warning about non-convergence and provides a diagnostic for whether to proceed with the fit. First, it is possible that this is not the correct model. With both the random block term and the spatial covariance structure, there may be too many terms in the model. There is some evidence of this from the "Random Effects Covariance Parameter Estimates" table, which shows a zero estimate for the block variance, Rep. However, the Convergence Score Test is as non-significant as can be, so perhaps we can still use the model.

Looking further to the "Repeated Effects Covariance Parameter Estimates" table, the estimate of the Spatial Spherical parameter, which is equivalent to the range in this model, is 714. Comparing this value to the variogram, that estimate is pretty clearly nonsense when it looks like the range should be between 20 and 30. The variogram does indicate that a nugget parameter is likely required, as there appears to be a jump in variance at the smallest distance.

Figure 7.7: Alliance Wheat Trial RCB with Spatial Structure Model Fit

Alliance - Fit Mixed

Fit Mixed

Fit Statistics

-2 Residual Log Likelihood	1074.8889
-2 Log Likelihood	1260.8772
AICc	1422.048
BIC	1580.1644

Warning: Parameter estimation algorithm did not converge.
Relative Gradient Criterion: 1.032

Convergence Score Test

ChiSquare	DF	Prob>ChiSq
1.032	59	1.0000

Possible causes:

- Model is not appropriate: Fixed effects, random effects, or repeated components might be incorrect.
- Model is too complex for the size of the data set.
- Data might have problematic outliers
- Data might be highly unbalanced
- Linear or near-linear dependencies exist among the input columns
- Response is skewed, highly discrete, or not approximately normally distributed

Possible solutions:

- If the convergence score test p-value is >.05, consider using the model and results with caution.
- Consider an alternative model, possibly a simpler one
- Remove outliers and rare levels of categorical factors
- Transform response so that it is more nearly normally distributed
- Increase the maximum iterations

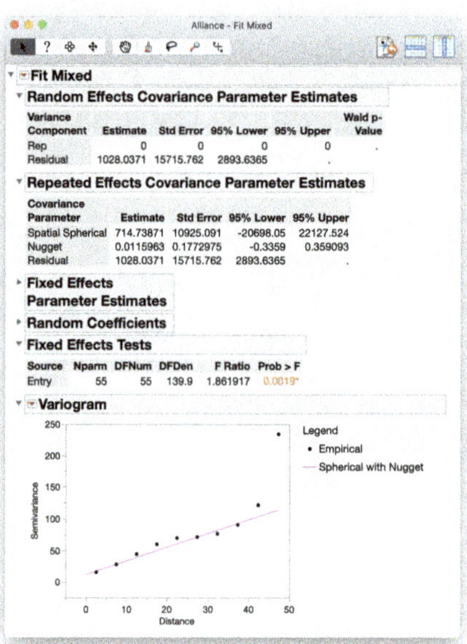

Alliance - Fit Mixed

Fit Mixed

Random Effects Covariance Parameter Estimates

Variance Component	Estimate	Std Error	95% Lower	95% Upper	Wald p-Value
Rep	0	0	0	0	.
Residual	1028.0371	15715.762	2893.6365	.	

Repeated Effects Covariance Parameter Estimates

Covariance Parameter	Estimate	Std Error	95% Lower	95% Upper
Spatial Spherical	714.73871	10925.091	-20698.05	22127.524
Nugget	0.0115963	0.1772975	-0.3359	0.359093
Residual	1028.0371	15715.762	2893.6365	.

Fixed Effects Parameter Estimates

Random Coefficients

Fixed Effects Tests

Source	Nparm	DFNum	DFDen	F Ratio	Prob > F
Entry	55	55	139.9	1.861917	0.0019*

Variogram

If the random block term is removed from the model, the algorithm still does not converge and yields similar spatial covariance parameter estimates to the model with the block included. In the interest of space, these results are not shown here. This is fairly common with spatial structures due to suboptimal starting values being chosen. We can try another spatial structure to see whether there is better behavior with it.

Spatial Model Instructions for Alliance Wheat Trial

Returning to the Model dialog box, on the *Random Effects* tab, select `Rep` and then click *Remove* to remove that term from the model.
On the *Repeated Structure* tab, change the Type to *Gaussian*.
Click *Run* to fit the model.

If you closed the Model dialog box, the complete instructions for this model follow. With the `Alliance.jmp` data table open, go to *Analyze > Fit Model*. Enter `Yield` as the *Y* variable. Change the default personality from *Standard Least Squares* to *Mixed Model*.
Select `Entry` in the *Select Columns* box then click *Add* to add fixed effect of variety to the *Construct Model Effects* box.
In the *Repeated Structure* tab of the *Construct Model Effects* box choose *Spatial with Nugget* from the Structure drop-down box. Choose *Gaussian* from the Type drop-down box. Enter `Latitude` and `Longitude` in the *Repeated* box.
Click *Run* to fit the model.

Results and Interpretation

The Gaussian spatial structure fit appears to model the variogram effectively. The "Repeated Effects Covariance Parameter Estimates" table in Figure 7.8 shows the range parameter, Spatial Gaussian, as 10.7. This corresponds to a practical range of about 18.5, which is reasonable from the variogram. The Nugget is 0.354, which needs to be scaled by the Residual for a full nugget value of 15.4. This also appears appropriate. Finally, the Residual parameter, which is equivalent to the geostatistical *partial sill* in this model with a nugget, is 43.4. The full sill value is partial sill + nugget, $43.4 + 15.4 = 58.8$; this value also looks reasonable given the variogram plateau.

Because we have reasonably modeled the spatial correlation, any conclusions regarding the fixed effects of varieties will now be more accurate. Compared to the RCB analysis in Figure 7.6, the Fixed Effects Test for `Entry` is now statistically significant, *p*-value = 0.0028, which more accurately reflects the observed differences in the field prior to harvest. As in the RCB analysis, we can add the LS Means Estimates for the varieties using the *Multiple Comparisons* option in the red triangle menu.

Figure 7.8: Alliance Wheat Trial with Gaussian Spatial Structure Model Fit

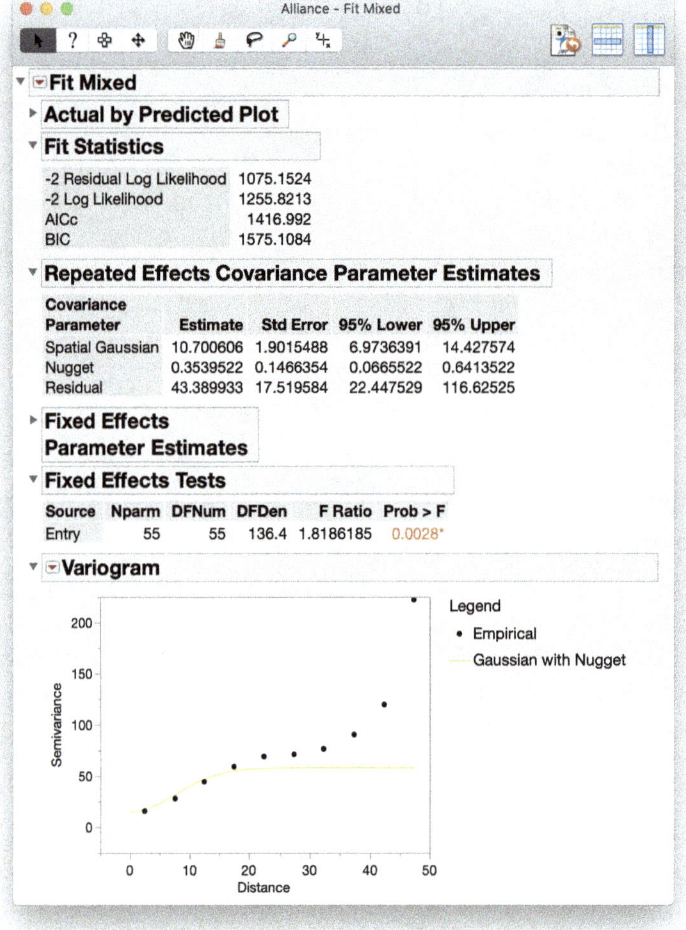

Figure 7.9 shows the LS Means Estimates sorted by Estimate value. After adjusting for the spatial correlation, the variety BUCKSKIN has the highest estimated yield of 35.6 compared with 25.6 in the RCB analysis. This also accurately reflects what the subject matter expert observed in the field. Further comparisons of interest could be done using the Multiple Comparison option.

Figure 7.9: Alliance Wheat Trial LSMeans from Gaussian Spatial Structure Model Fit

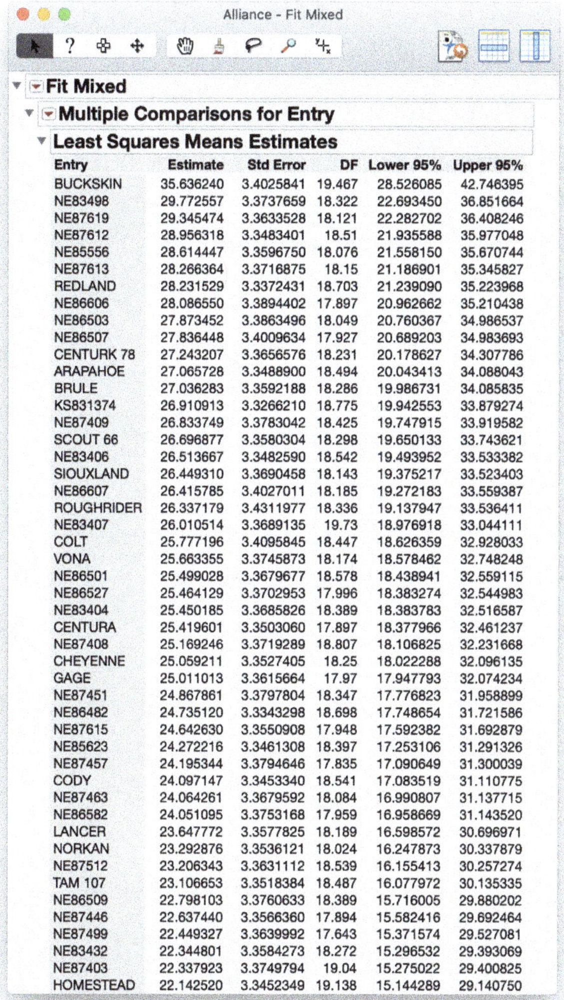

Alliance - Fit Mixed

Fit Mixed

Multiple Comparisons for Entry

Least Squares Means Estimates

Entry	Estimate	Std Error	DF	Lower 95%	Upper 95%
BUCKSKIN	35.636240	3.4025841	19.467	28.526085	42.746395
NE83498	29.772557	3.3737659	18.322	22.693450	36.851664
NE87619	29.345474	3.3633528	18.121	22.282702	36.408246
NE87612	28.956318	3.3483401	18.51	21.935588	35.977048
NE85556	28.614447	3.3596750	18.076	21.558150	35.670744
NE87613	28.266364	3.3716875	18.15	21.186901	35.345827
REDLAND	28.231529	3.3372431	18.703	21.239090	35.223968
NE86606	28.086550	3.3894402	17.897	20.962662	35.210438
NE86503	27.873452	3.3863496	18.049	20.760367	34.986537
NE86507	27.836448	3.4009634	17.927	20.689203	34.983693
CENTURK 78	27.243207	3.3656576	18.231	20.178627	34.307786
ARAPAHOE	27.065728	3.3488900	18.494	20.043413	34.088043
BRULE	27.036283	3.3592188	18.286	19.986731	34.085835
KS831374	26.910913	3.3266210	18.775	19.942553	33.879274
NE87409	26.833749	3.3783042	18.425	19.747915	33.919582
SCOUT 66	26.696877	3.3580304	18.298	19.650133	33.743621
NE83406	26.513667	3.3482590	18.542	19.493952	33.533382
SIOUXLAND	26.449310	3.3690458	18.143	19.375217	33.523403
NE86607	26.415785	3.4027011	18.185	19.272183	33.559387
ROUGHRIDER	26.337179	3.4311977	18.336	19.137947	33.536411
NE83407	26.010514	3.3689135	19.73	18.976918	33.044111
COLT	25.777196	3.4095845	18.447	18.626359	32.928033
VONA	25.663355	3.3745873	18.174	18.578462	32.748248
NE86501	25.499028	3.3679677	18.578	18.438941	32.559115
NE86527	25.464129	3.3702953	17.996	18.383274	32.544983
NE83404	25.450185	3.3685826	18.389	18.383783	32.516587
CENTURA	25.419601	3.3503060	17.897	18.377966	32.461237
NE87408	25.169246	3.3719289	18.807	18.106825	32.231668
CHEYENNE	25.059211	3.3527405	18.25	18.022288	32.096135
GAGE	25.011013	3.3615664	17.97	17.947793	32.074234
NE87451	24.867861	3.3797804	18.347	17.776823	31.958899
NE86482	24.735120	3.3343298	18.698	17.748654	31.721586
NE87615	24.642630	3.3550908	17.948	17.592382	31.692879
NE85623	24.272216	3.3461308	18.397	17.253106	31.291326
NE87457	24.195344	3.3794646	17.835	17.090649	31.300039
CODY	24.097147	3.3453340	18.541	17.083519	31.110775
NE87463	24.064261	3.3679592	18.084	16.990807	31.137715
NE86582	24.051095	3.3753168	17.959	16.958669	31.143520
LANCER	23.647772	3.3577825	18.189	16.598572	30.696971
NORKAN	23.292876	3.3536121	18.024	16.247873	30.337879
NE87512	23.206343	3.3631112	18.539	16.155413	30.257274
TAM 107	23.106653	3.3518384	18.487	16.077972	30.135335
NE86509	22.798103	3.3760633	18.389	15.716005	29.880202
NE87446	22.637440	3.3566360	17.894	15.582416	29.692464
NE87499	22.449327	3.3639992	17.643	15.371574	29.527081
NE83432	22.344801	3.3584273	18.272	15.296532	29.393069
NE87403	22.337923	3.3749794	19.04	15.275022	29.400825
HOMESTEAD	22.142520	3.3452349	19.138	15.144289	29.140750

This example showed how a spatial analysis "saved" the time and effort involved in a large-scale trial where the experiment design was not ideal. Spatial models are not a substitute for good experiment design, however. An incomplete block design with appropriate block sizes for this field would likely be more effective. Stroup (2002) shows that spatial analyses on data like this example are relatively inefficient compared to incomplete block designs. As always, good experiment design is essential to effective experimentation.

7.5 Further Statistical Details

JMP Pro includes four spatial covariance structures with and without nugget effects. They are power, exponential, Gaussian, and spherical. The formulas for $f(d_{ij})$ for these are as follows.
Power: $f(d_{ij}) = \rho^{d_{ij}}$
Exponential: $f(d_{ij}) = exp(-d_{ij}/\rho)$
Gaussian: $f(d_{ij}) = exp(-d_{ij}^2/\rho^2)$
Spherical: $f(d_{ij}) = [1 - 1.5(d_{ij}/\rho) + 0.5(d_{ij}/\rho)^3 \times 1(d_{ij} < \rho)$

$1(d_{ij} < \rho)$ is an indicator function that equals 1 when $d_{ij} < \rho$ and 0 otherwise.

ρ is the range parameter, but it only equals the range in the spherical structure. The other three structures only reach zero covariance asymptotically. Therefore the *practical range* is the distance where the variogram reaches 95% of the sill. For exponential models, the practical range is 3ρ. For Gaussian, it is $\rho\sqrt{3}$.

Table 7.2: Geostatistical Terminology

JMP Covariance Parameter Name	Geostatistical Term	Scaled Y/N
Spatial [Structure Name]	Range Parameter	Yes, per practical range above
Nugget	Nugget	Yes, multiply by Residual for full Nugget
Residual	Sill or Partial Sill	No

7.6 Exercises

The data for these exercises can be found in `Seed Trial.jmp`. The experiment was similar to the Alliance wheat trial of the second example. The variable **`Rep`** is the complete block with 3 complete blocks of the 48 treatments, **`Trt`**. The variable **`Block`** divides the complete blocks into incomplete blocks for a BIB design. The location of each experimental unit in the 12×12 grid was also recorded as **`Lat`** and **`Lng`**. The response was yield and recorded as **`Y`**. Higher yield is desirable.

1. Fit an RCBD model to these data. Why might the RCB not be appropriate in this case? Is there any statistical output from fitting the model that indicate a problem? If so, what?
2. Using the spatial information in the data table, identify an appropriate spatial covariance structure for these data. Is there a nugget effect present? What evidence led you to choose the structure that you selected?
3. Fit the BIB model to these data. Does this model seem appropriate? Why or why not?

4. Statistics is sometimes referred to as the "art and science of learning from data." Part of the 'art' is selecting and defending the model that you chose to analyze the data. Which of the models fit in these exercises do you choose to use to complete the analysis? Why (i.e. best fit statistic, smallest standard error of treatments or treatment differences, most appropriate given the design of the experiment)?
5. Using the model you that selected, complete the analysis identifying the 'best' treatments to move forward into further research and development. Would these 'best' treatments have been different if you had used a different model?

Chapter 8

Simulation and Power Analysis

We have left discussion of sample size and power to later in the book, but in reality these calculations should be done at the design phase of the experiment, long before analyzing data. If the experiment that you run is not going to have the power to detect the difference you are hoping to see, then all of the time and materials used in the experiment are potentially wasted. Power and sample size calculators are available in many software packages, including JMP, but most of these options do not take mixed model analyses into consideration in their calculations. When there is more than one source of variability, tests (and confidence intervals) for means and their differences often have complex combinations of the variance components in the standard error. This makes "out of the box" calculators using only the residual variance inappropriate for those tests and can lead to overestimating power.

We will focus on using simulation to estimate power and confidence interval coverage in this chapter. Simulation is a powerful way to understand how any statistical procedure performs under different circumstances. You can simulate data in many different ways, and when repeating that design many times, see how the statistics change (or, better, do not!) with each simulated data set. Most of the examples in this chapter use the Simulate feature available in JMP Pro. The final example using scripting does not require JMP Pro, though the script could be modified to use models only available in JMP Pro.

8.1 Motivating Examples

Semiconductor Experiment — Semiconductor researchers have process conditions and finishing treatments that can interact in affecting performance of the chip when complete. They have a new design that they need to investigate how it will perform given what they already know about their chip manufacturing process.

Wheat Trial — The wheat farmers need to know whether the statistical model that they are using to analyze their data produces confidence intervals that are actually 95% intervals as they state. They have heard that sometimes a procedure only has coverage of 90% or less despite stating 95%. They want to be confident about their confidence.

All the Statistics — A researcher wants to investigate the properties of several different statistics produced in analyzing a randomized complete block design. They would like a compact way to study them all without having to run separate studies for each individual statistic.

8.2 Simulation for Precision and Power

Simulation studies are a way of examining the behavior of a statistical procedure using different experiment design conditions. JMP Pro includes a powerful Simulate option that makes performing this sort of study straightforward. The idea of a simulation study is to fit a model many times with randomly generated data using parameter values of interest each time and accumulate the statistic of interest. That statistic could be a p-value to estimate power through the number of rejections of the null hypothesis or the standard error to estimate the width of a confidence interval. To create the randomly generated data, you add a simulation formula column to the data table. This simulation formula column is then used as the Y response variable in each model fit.

There are two built-in methods for creating this simulated response column. The first uses the Design of Experiments Custom Design platform in JMP. The second uses a Save Column option within the Fit Model platform's Mixed personality. The DOE option works well for CRD, RCBD, and split-plot designs and makes it easy to specify variance component values. The Mixed option works with any non-repeated measures model, including complicated random coefficient models. It is somewhat more difficult to modify the variance component values as a result of the complicated models.

DOE Simulate

An option in the Custom Design platform enables you to simulate responses given the design that you have created. The Custom Design platform is a powerful tool to create many different types of experiment designs. In this example, we are going to keep it simple with a split-plot design similar to the semiconductor experiment in Chapter 3. In this case, the process condition is applied to the entire wafer, then after splitting the wafer into the smaller chips a finishing treatment is applied. For simplicity, only two levels each of the process condition and finishing treatment are considered.

Creating a Split-Plot Design in Custom Design and Simulating Responses

From any JMP window select *DOE > Custom Design* to launch the Custom Design platform.
In the Factors outline box, type 2 next to *Add N Factors*.
Click *Add Factor > Categorical > 2 Level*.
If it is not already selected, double-click on `X1` to select it and rename it `Process`.
Double-click on or Tab to select `X2` to rename it `Finish`.
The process condition is the hard-to-change, whole plot factor. Next to `Process` click *Easy* and select *Hard*. This defines `Process` as the whole plot factor and `Finish` as the split-plot factor.
Click *Continue*.

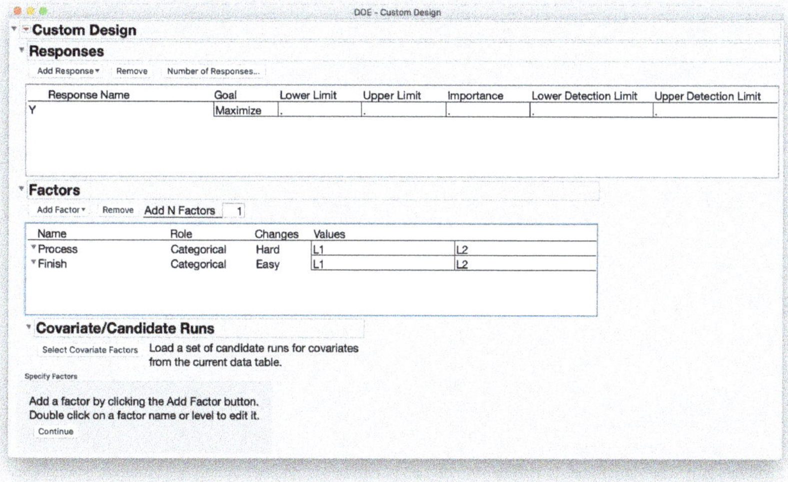

DOE - Custom Design

Custom Design

Responses

Add Response ▾ Remove Number of Responses...

Response Name	Goal	Lower Limit	Upper Limit	Importance	Lower Detection Limit	Upper Detection Limit
Y	Maximize	.	.	.	.	.

Factors

Add Factor ▾ Remove Add N Factors 1

Name	Role	Changes	Values	
Process	Categorical	Hard	L1	L2
Finish	Categorical	Easy	L1	L2

Covariate/Candidate Runs

Select Covariate Factors Load a set of candidate runs for covariates from the current data table.

Specify Factors

Add a factor by clicking the Add Factor button.
Double click on a factor name or level to edit it.
Continue

Creating a Split-Plot Design and Simulating Responses, Continued

In the Model outline box, select *Interactions* > *2nd*. This adds the important two way interaction to the factorial model.

The Custom Design platform sets a default Number of Whole Plots to 4 and a default Number of Runs to 12. That implies 3 split-plots per whole plot, but we only have two split-plot factor levels, and the wafers are split into two pieces. To make it even, we can either increase the number of whole plots to 6 or decrease the number of runs to 8. Which to choose is part of the power and sample size problem that we are solving and also has cost considerations. Sometimes resources are limited, so we need to restrict the number of runs. But if the cost difference between 8 and 12 is acceptable, more runs will give more power. Let's say our resources are limited and set the *User Specified* number of runs to 8.

Click the Custom Design red triangle and select *Simulate Responses*. This command will open the Simulate Responses window after the design table is constructed.

If you want to replicate the design exactly in the future, you need to specify the number of starts for the design algorithm and a random seed. Click the Custom Design red triangle and select *Set Random Seed*. Then enter the seed of your choice. In this example, we used 68310. Click the Custom Design red triangle and select *Number of Starts*. Then enter the number. We chose 2500.

The screenshot below shows the completed Custom Design dialog box prior to making the design. Click *Make Design* to make the design. Click *Make Table* to create the data table with the experiment design and simulation column.

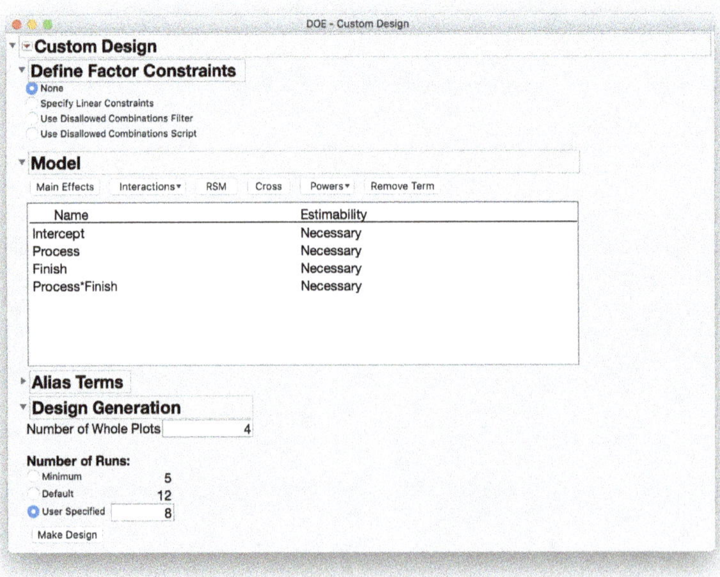

Figure 8.1 shows the resulting experiment design data table and the Simulate Responses box. The data table includes the design information, whole plots, process and finish levels assigned to each run, a **Y** column which after completing the experiment can hold your observed response values, and a **Y Simulated** column. Use the Simulate Responses dialog window to specify the model parameters for your simulation. The model parameters include the effects and the variance components. If your simulation study goal is to determine the power of the design, set the effect parameters to values that represent the "minimum relevant difference" you hope to detect in your study. With the effect coding, the coefficients that you enter will be one-half the relevant difference (for example, 5 if your relevant difference is 10) as the last level will be set to sum-to-zero. This is demonstrated in detail in Section 8.4. If you are studying the Type I error characteristics, also referred to as the size of the test, you will set the effect values equal to zero. Finally, using either subject-matter expertise or a pilot study, you set the variance component values. In this case, the dialog box takes the standard deviation rather than the variance.

Figure 8.1: Custom Design Table and Simulation Dialog Box

Custom Design

Custom Design
Design Custom Design
Criterion A-Optimal
Model
Evaluate Design
Fit Mixed
DOE Simulate
DOE Dialog

Columns (5/0)
Whole Plots *
Process *
Finish *
Y *
Y Simulated + *

Rows
All rows 8
Selected 0
Excluded 0
Hidden 0
Labeled 0

	Whole Plots	Process	Finish	Y	Y Simulated
1	1	L1	L1	2.4216200...	2.42162008
2	1	L1	L2	-1.045557...	-1.0455573
3	2	L1	L2	1.0756058...	1.0756058
4	2	L1	L1	5.5870057...	5.58700576
5	3	L2	L1	2.9394056...	2.93940566
6	3	L2	L2	0.0010220...	0.00102209
7	4	L2	L2	0.2689004...	0.26890048
8	4	L2	L1	0.1554361...	0.15543618

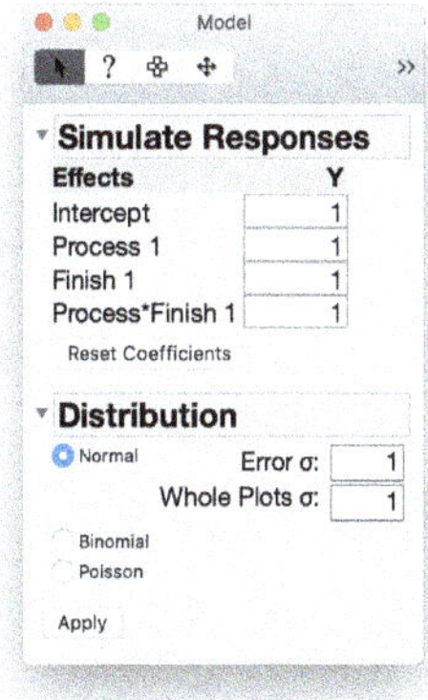

Creating a Split-Plot Design in Custom Design and Simulating Responses

We are going to use this design to investigate the size of the test for the main effects and interactions in the example later in this chapter, so these instructions set up the simulated response column for that study.

In the Simulated Responses window, change the values next to all of the Effects to 0. Because this dialog box uses the effect coding structure, if you set the values equal to a nonzero value, the last level of the effect (notice it is not listed in the dialog box) will be set to -(sum of other effects). That value is clearly not equal to the others unless the others are all zero.

We have the Semiconductor Experiment analysis that can serve as a pilot study for this design's variance components. In that analysis the whole plot variance was $\sigma_w^2 = 0.1057903$, and the residual variance was $\sigma^2 = 0.1111493$. The Random Normal function in JMP requires the standard deviation rather than variance, so we take the square root of these values to enter in the dialog box. The Whole Plot σ is approximately 0.33. The Error σ is also approximately 0.33, but we will round up to 0.34 to ensure that the residual variance is *at least* as large as in the pilot study.

Click *Apply* to apply these changes to the `Y Simulated` formula in the Custom Design data table. This sets the values from the defaults to the specified values to simulate the responses.

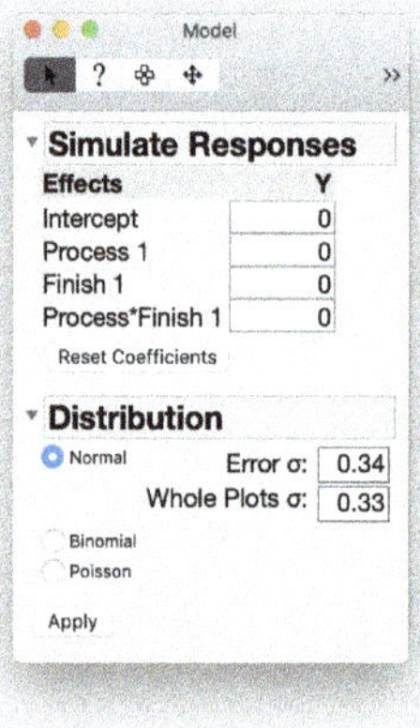

The final nice feature of using the Custom Design platform to create our design and simulation formula is that the resulting data table includes a *Fit Mixed* script to fit the appropriate model. We will use this script when we complete the simulation study example in Section 8.3.

Fit Mixed Simulate

To demonstrate the simulation formula in the Fit Mixed platform, we will use the **winterwheat.jmp** data referenced in the Chapter 5 exercises. This feature requires JMP Pro 16. In this case, we are going to use that data as a pilot study to use for further power studies. We begin by fitting the random coefficient model.

Random Coefficient Instructions for Winter Wheat

With the **winterwheat.jmp** data table open, go to *Analyze > Fit Model*. Enter **yield** as the *Y* variable. Change the default personality from *Standard Least Squares* to *Mixed Model*.

Select **moist** in the *Select Columns* box then click *Add* to add fixed effect of moisture to the *Construct Model Effects* box.

Enter **moist** in the *Random Effects* tab of the *Construct Model Effects* box. Then select **variety** in the *Select Columns* box and **moist** in the *Random Effects* tab and click *Nest Random Coefficients* to create the batch-specific random intercept and slope.

Click *Run* to fit the model.

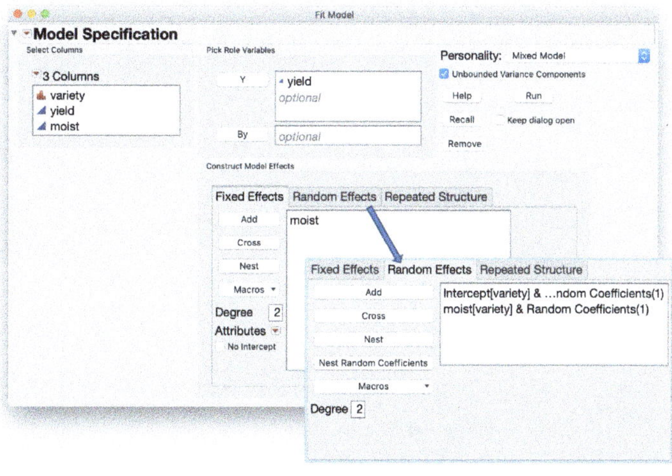

Once the Fit Mixed report has run, select the *Save Simulation Formula* option from the *Save Columns* submenu of the red triangle menu. This adds a column **yield Simulation Formula** to the data table. This is the column to "switch in" when using the Simulate feature.

Details of the `yield Simulation Formula`

The key difference between the formula created by the DOE simulation formula option and the Fit Mixed simulation formula is in the final generation of the response value. In the DOE formula, we saw that it follows the statistical model format with the whole-plot effect, the split-plot effect, the fixed effects, and the residual effect being summed. The Fit Mixed formula uses the assumption that the vector of responses $\mathbf{Y}$ is distributed multivariate normal with mean $\boldsymbol{\mu}$ and variance $\mathbf{ZGZ}'+\mathbf{R}$. This enables the more complex G-side covariance structures in the random coefficient model to be used.

In this formula, the parameters most easily changed are the fixed effect parameters stored in the formula vector `_Beta`. The variance component parameters are embedded in the G and R matrices. They can be modified, but care needs to be taken to be sure the structure of the matrices remains intact. Also, because JMP creates and stores these matrices, the simulation process can run very slowly if the data table is large. How large is too large depends on the computer and user patience. Because most pilot studies are relatively small, this should not be a large barrier.

The only values that change in the simulations are the `_Y` values as each simulate run calls the `Random Normal` function. Every simulation run uses the same parameters to generate the responses.

```
If(
  Row()==1 ⇒ _X=[60 × 2];
             _Beta=[33.43
                    0.662];
             _Mu=_X•_Beta;
             _Z=[60 × 20];
             _G=[20 × 20];
             _R=[60 × 60];
             _V=_Z•_G•_Z`+_R;
             _lambda=Cholesky(_V);
             _Y=_Mu+_lambda•J(N Row(_lambda),1,Random Normal(<mu=0>));
  else       ⇒ else clause
);
_Y Row();
```

Using the Simulate Formula for Power

When you have your simulation formula column from either method, you can now simulate results for any statistic of interest in the JMP report. Which statistic you choose to focus your simulation on will depend on your goals. For a power analysis, a probability column in the report is used. Parametric bootstrap confidence intervals are obtained when using an Estimate column. To study confidence interval coverage, a few additional steps are needed after using the confidence interval columns. We will look at examples of each of these after a quick explanation of the Simulate function.

Simulate Instructions

With the **`winterwheat.jmp`** random coefficient model report run, right-click on the *Prob > F* column in the Fixed Effects Tests section of the report then choose *Simulate*.
A dialog box appears with the options for the simulation. The *Column to Switch Out* contains a list of the columns that were used in the original report. By default it chooses the column that was used as the *Y*, response, column, which is what we want for this simulation. The *Column to Switch In* contains a list of all columns in the data table that include a random function in it because we need new random data with each run. In this case, we only have the **`yield Simulation Formula`** column, so it is selected. The *Number of Samples* box enables you to set how large a simulation you want to run. Using a small number, like 10, enables you to see how long it takes to perform a larger simulation and whether the results are as you expect. When you know you will get what you want, a larger number, typically 1000 or more, should be used. Finally, you can set a *Random Seed* to replicate the results in future simulations exactly.
Once the options are set, click *OK* to run the simulation.

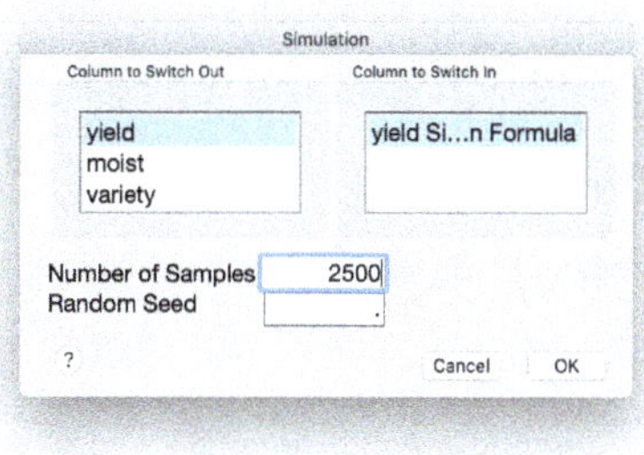

After clicking *OK*, JMP then fits the model that had been specified for the initial report now using the simulated data. Behind the scenes, the statistic of interest is being accumulated from each sample fit. Depending on the size of the simulation study, a progress bar will appear. When the simulations are complete, a new data table with the results is displayed. This data table is shown in Figure 8.2. The table contains a column referencing the original data table, a column with the ID of the simulation (from 0, the original statistic to the number of samples requested), and a column (or columns, as we will see in the detailed examples) with the statistic calculated in that simulation, in this

Figure 8.2: Simulation Results Data Table

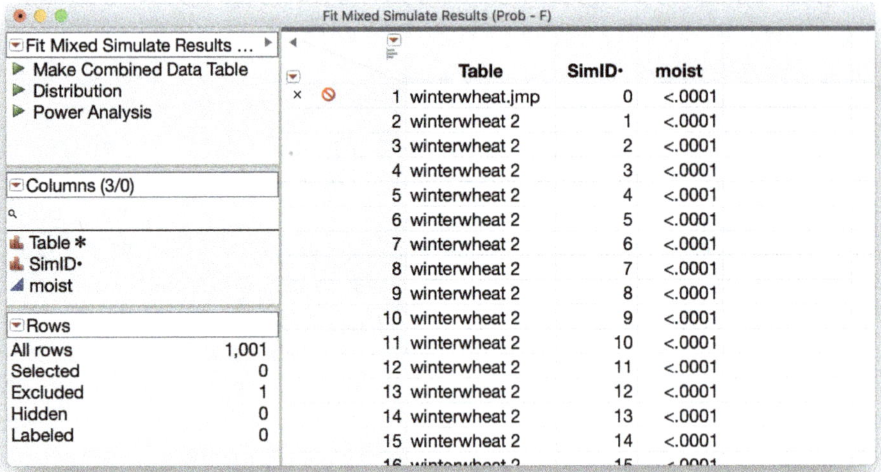

Fit Mixed Simulate Results (Prob - F)

Fit Mixed Simulate Results ...
- Make Combined Data Table
- Distribution
- Power Analysis

Columns (3/0)
- Table *
- SimID•
- moist

Rows	
All rows	1,001
Selected	0
Excluded	1
Hidden	0
Labeled	0

	Table	SimID•	moist
1	winterwheat.jmp	0	<.0001
2	winterwheat 2	1	<.0001
3	winterwheat 2	2	<.0001
4	winterwheat 2	3	<.0001
5	winterwheat 2	4	<.0001
6	winterwheat 2	5	<.0001
7	winterwheat 2	6	<.0001
8	winterwheat 2	7	<.0001
9	winterwheat 2	8	<.0001
10	winterwheat 2	9	<.0001
11	winterwheat 2	10	<.0001
12	winterwheat 2	11	<.0001
13	winterwheat 2	12	<.0001
14	winterwheat 2	13	<.0001
15	winterwheat 2	14	<.0001

case the `moist` *p*-value. JMP also includes one or two scripts for completing the analysis. The Distribution script is always provided and will produce a *Distribution* report of the statistic column(s). A Power Analysis script is included if the statistic column is a *p*-value, anticipating the desire to test the power performance. We will discuss the results of those scripts in conjunction with the full examples.

Simulation is as easy as that with this powerful tool built into JMP Pro!

8.3 Type I Error Control Using the Semiconductor Design

An experiment design that does not control Type I error should not be considered for use. Any power calculation will be spurious due to over- (or under-) rejection of the null hypothesis. When considering a new design, an examination of its Type I error control should be done first. If the design maintains the proper size of the test (i.e., alpha-level), then further investigation of the power characteristics can be done. We will investigate the Type I error control of the split-plot design of the semiconductor experiment. To verify control, we simulate the data many times, collect the *p*-value used to determine rejection of the null hypothesis, and count how many times the null hypothesis is rejected at a specified alpha level. If the proportion of rejections is approximately equal to the level of alpha, the Type I error is as expected.

Simulate Instructions

Begin with the `Custom Design.jmp` data table that was created in Section 8.2. If you do not have that data table, follow the instructions in that example, including setting the effects and variance components, then return to these steps.

The data table created by the *Custom Design* platform includes a Fit Mixed script that contains the proper model for the generated design. Click on the green triangle next to Fit Mixed in the data table to run the *Mixed* platform.

Now we have the Fit Mixed report from which to simulate. Right-click on the *Prob > F* column of the Fixed Effects Tests section of the report and select *Simulate*.

The Fit Mixed script ran using the `Y` column as the response, so we want to switch it out with the `Y Simulated` column for the simulation.

Enter how many simulations you want to run in the *Number of Samples* box. Here we have done 1000.

If you want to replicate your results exactly in the future, set a *Random Seed*. To replicate our results, enter 582397.

Once the options are set, click *OK* to run the simulation.

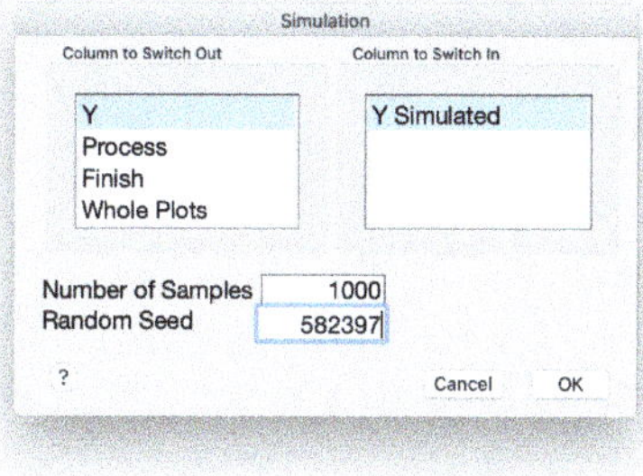

Figure 8.3 shows the data table of results of the simulation. It includes columns for each of the effects in the model, Finish, Process, and their interaction. We can see just glancing at the table that most of the resulting p-values would be considered non-significant, which is correct given the parameters of the simulated data. We can use either the *Distribution* or *Power Analysis* scripts included in the data table to quickly summarize the results. The Power Analysis script streamlines the report to the most salient sections of the distributions of the simulation p-values. Click on the green arrow next to *Power Analysis* to run the script.

Figure 8.3: Simulation Results Data Table

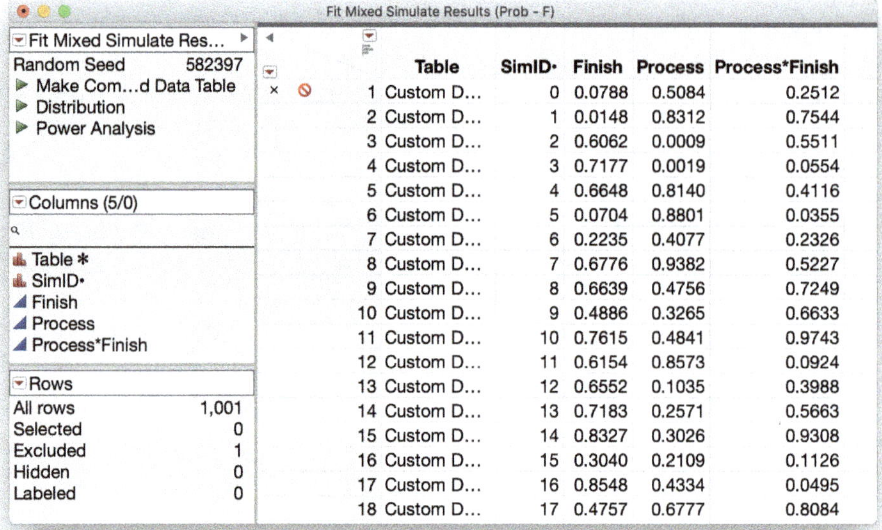

Fit Mixed Simulate Results (Prob - F)

Fit Mixed Simulate Res...
Random Seed 582397
Make Com...d Data Table
Distribution
Power Analysis

Columns (5/0)
Table
SimID•
Finish
Process
Process*Finish

Rows
All rows 1,001
Selected 0
Excluded 1
Hidden 0
Labeled 0

	Table	SimID•	Finish	Process	Process*Finish
1	Custom D...	0	0.0788	0.5084	0.2512
2	Custom D...	1	0.0148	0.8312	0.7544
3	Custom D...	2	0.6062	0.0009	0.5511
4	Custom D...	3	0.7177	0.0019	0.0554
5	Custom D...	4	0.6648	0.8140	0.4116
6	Custom D...	5	0.0704	0.8801	0.0355
7	Custom D...	6	0.2235	0.4077	0.2326
8	Custom D...	7	0.6776	0.9382	0.5227
9	Custom D...	8	0.6639	0.4756	0.7249
10	Custom D...	9	0.4886	0.3265	0.6633
11	Custom D...	10	0.7615	0.4841	0.9743
12	Custom D...	11	0.6154	0.8573	0.0924
13	Custom D...	12	0.6552	0.1035	0.3988
14	Custom D...	13	0.7183	0.2571	0.5663
15	Custom D...	14	0.8327	0.3026	0.9308
16	Custom D...	15	0.3040	0.2109	0.1126
17	Custom D...	16	0.8548	0.4334	0.0495
18	Custom D...	17	0.4757	0.6777	0.8084

Figure 8.4 shows the Distribution report of the p-values for the three model effects. Each section for the effects begins with a histogram of the p-values. Because the parameters for the effects were set to zero, we would expect an approximately uniform distribution of values between 0 and 1. The histograms appear uniform. The second section, Simulation Results, displays the original estimate of the p-value (labeled SimID 0 in the data table) for reference and includes bootstrap confidence intervals at various confidence levels and empirical p-values for tests of whether there is a difference from the original estimate. In a simulation investigating the Type I error characteristics, this section is not particularly useful.

The final section, Simulated Power, shows the main results of interest. When testing a hypothesis at a specified alpha-level, we are stating that if the p-value is smaller than the stated alpha the result is unlikely *if the null hypothesis is true*. That does not mean that we would never see such a result; unlikely events happen quite regularly. By setting the alpha-level at $\alpha = 0.05$, we are stating that we could be wrong 5 times out of 100, or relevant to this simulation, 50 out of 1000. Therefore, when testing the Type I characteristics of a test, we *expect* to see rejections of the null hypothesis at a rate approximately equal to the alpha that we have chosen.

Here we see for **`Finish`**, the observed Rejection Rates are all very close to their respective Alphas. For `Process` the Rejection Rates are a little lower, but they are still within the expected range for a simulation study of this size. The results for the `Process*Finish` interaction are again consistent with their alpha-levels. We can conclude from this study that this split-plot design maintains its size, and further power studies will not be impacted by over- or under-rejection.

8.4 Power Using the Semiconductor Design

In the previous section, we validated the Type I error characteristics of the split-plot design used for the Semiconductor experiment. With that confirmation, we can proceed to investigate the power of the design to detect various differences at specified alpha-levels. The scientists know that an improvement in the response of 0.5 units is great, and an improvement of 0.25 units is good and would be nice to detect. The researchers also believe there is an interaction between the process condition and finish treatment and would like to have the power to detect this interaction.

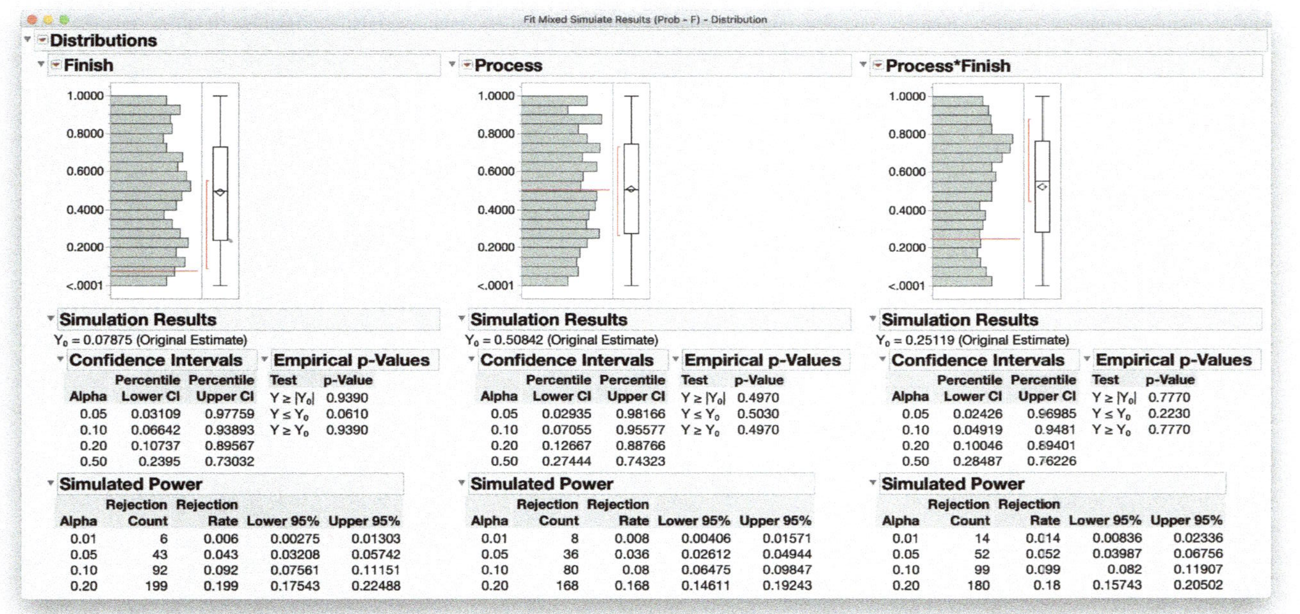

Figure 8.4: Power Analysis of Simulation Results

Power Simulation Instructions - Setting Parameters

Begin with the **Custom Design.jmp** data table that was created in Section 8.2. If you do not have that data table, follow the instructions in that example, up to setting the effects and variance components, then return to these steps.
You can modify the simulation formula in the Custom Design data table using the same Simulate Responses dialog as before. If it is no longer open, click the green triangle next to the *DOE Simulate* script to run that script. It will restart the simulate process.
In the Simulate Responses dialog box, enter 0 for the Intercept effect. For the Process 1 effect enter 0. This implies no main effect of Process, which may not be true. However, with the belief of an interaction, a main effect analysis is inappropriate, and we should focus on the simple effects. Enter 0.125 for the Finish 1 effect. This will correspond to an 0.25 unit change in the response as a result of the effect coding. The formula will set Finish 1 = 0.125 and Finish 2 = -0.125 for an 0.25 difference. Finally, enter 0.25 for the **Process*Finish 1** effect. The effect coding process will automatically set up the remaining **Process*Finish** effects to include the interaction effect of interest.
The variance components should be set the same as in our Type I simulation. In that case from our pilot study, the Error σ was 0.34 and the Whole Plots σ was 0.33.
With all of the effects coefficients and variance component parameters set, click *Apply* to update the **Y Simulated** column.

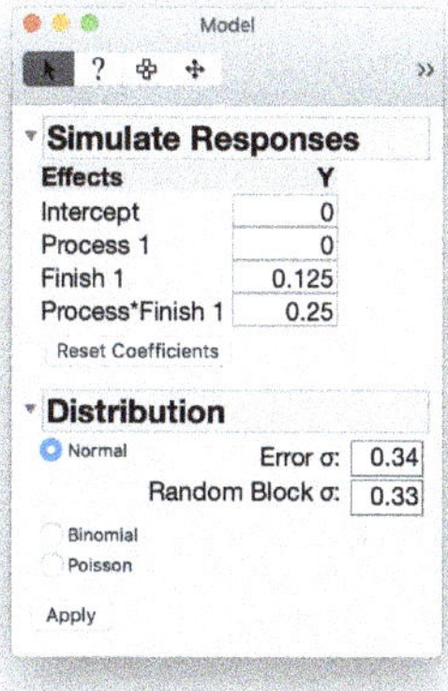

Power Simulation Instructions - Simulating

Click the green triangle next to *Fit Mixed* to run the mixed model script.
The first statistic of interest is the interaction fixed effect because having power to detect that interaction is key. Right-click on the *Prob > F* column of the Fixed Effects Tests section of the report and select *Simulate*.
The Fit Mixed script ran using the `Y` column as the response, so we want to switch it out with the `Y Simulated` column for the simulation.
Enter how many simulations you want to run in the *Number of Samples* box. Here we have done 1000.
If you want to replicate your results exactly in the future, set a *Random Seed*. To replicate our results, enter 347865.
Once the options are set, click *OK* to run the simulation.

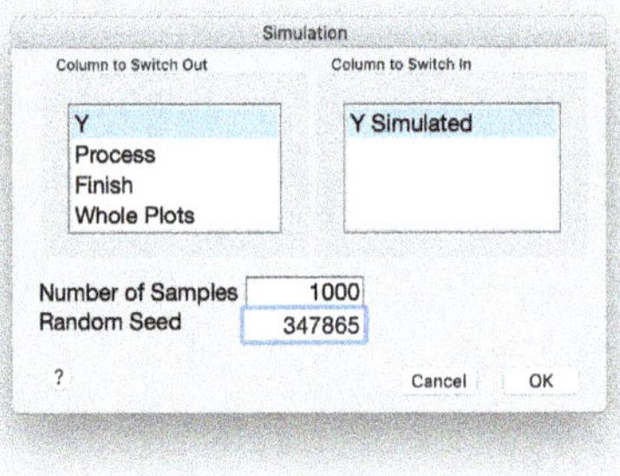

When the simulation is complete, the results data table will appear with the collected *p*-values from each simulation run. Scanning through the table, we notice a few more *p*-values that would be statistically significant than in the Type I simulation. At least in some runs, the interaction effect is "significant" as is the effect of Finish. We set the Process coefficient at 0, so we would expect not to see many small probabilities in the `Process` column. Click the green arrow next to *Power Analysis* to run the script to summarize the results.

We are interested in the results of the `Process*Finish` interaction effect, so Figure 8.5 shows the distribution report with the `Finish` and `Process` outline nodes closed. If we use the usual alpha-level of 0.05, the simulated power for this test is only 23.0%. However, analysts often use the more relaxed alpha of 0.20 for interaction tests because they would rather include an interaction when there is not one than omit an interaction that is present. For an $\alpha = 0.20$, the rejection rate is 64.6%. This may be sufficient to the researchers, but often studies are designed to achieve 80% power. If the semiconductor researchers want that level, more whole plot experimental units will be required for the study. We will do that at the end of this example.

Figure 8.5: Power Analysis of Simulation Results

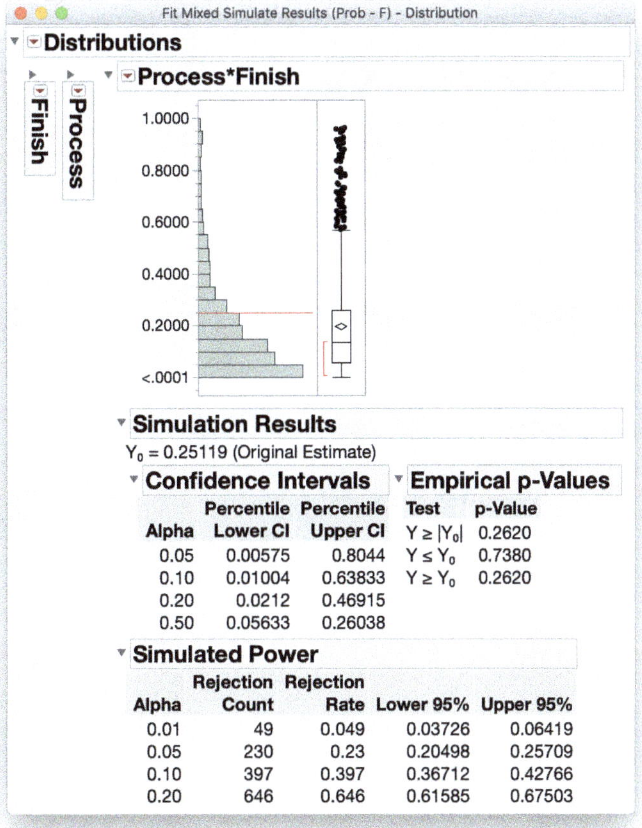

Alpha	Percentile Lower CI	Percentile Upper CI
0.05	0.00575	0.8044
0.10	0.01004	0.63833
0.20	0.0212	0.46915
0.50	0.05633	0.26038

Test	p-Value
$Y \geq \lvert Y_0 \rvert$	0.2620
$Y \leq Y_0$	0.7380
$Y \geq Y_0$	0.2620

Alpha	Rejection Count	Rejection Rate	Lower 95%	Upper 95%
0.01	49	0.049	0.03726	0.06419
0.05	230	0.23	0.20498	0.25709
0.10	397	0.397	0.36712	0.42766
0.20	646	0.646	0.61585	0.67503

The researchers were also interested in being able to detect as small as a 0.25 unit difference in the response. We can test the power for this using the Multiple Comparisons tool. In the Fit Mixed report, choose *Multiple Comparisons* from the red triangle menu. The `Process*Finish` effect is our effect of interest, so select it in the *Choose an Effect* list. Checking the box for *All Pairwise Comparisons - Student's t* will include the tests of differences of interest. We are pre-planning our difference tests, so we do not need to use a multiplicity adjustment. Then click *OK* to add the comparisons to the Fit Mixed report. Figure 8.6 shows the completed multiple comparisons dialog box.

Figure 8.6: Multiple Comparisons Dialog Box

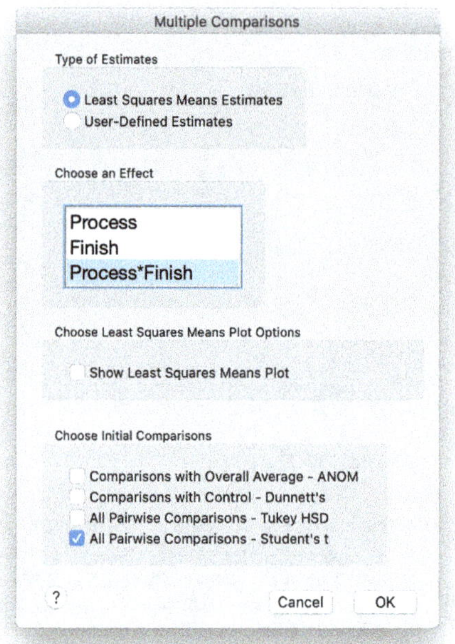

The simple effect pairwise comparisons include these planned differences.

Process Level	Finish Level	-Process Level	-Finish Level	Planned Difference
1	1	1	2	0.75
1	1	2	1	0.50
1	2	2	2	0.50
2	1	2	2	0.25

These are legitimate simple effects tests because one factor remains at the same level within the comparison. It also happens that the 0.75 and 0.25 comparisons occur within whole plots (the Process level remains constant), which decreases variability; with a smaller standard error, test statistics are larger and confidence intervals are more narrow. The smaller standard error will increase power compared to the same difference tested with a larger standard error.

Right-click on the *Prob* > |*t*| column in the All Pairwise Differences report and select *Simulate*. We again use 1000 simulations and set a seed of 763548 for this example. To help us remember which comparison is for what difference, we added the value to the name of the corresponding column in the data table before running the Power Analysis script. Figure 8.7 shows the power analysis with the comparisons of interest expanded, and those not of interest closed.

This design with only four whole plots may be able to detect the interaction, but it does not have very good power characteristics for the comparisons the researchers were interested in. Even the 0.75 unit difference would only have been declared 'significant' 23.9% of the time. The ideal 0.25 unit difference was only rejected 6.8% of the time, barely more than noise. To detect the 0.25 difference, a much larger experiment is required.

To increase the experiment design, we can go back to the DOE Custom Design platform. It is possible the platform dialog box is still open, if you have not closed it, as it remains open when creating the data table. If it is open, click the *Back* button to go back to the Design Generation phase. If it is not open, you can run the *DOE Dialog* script in the Custom Design table to reopen the platform dialog box. At this point, it becomes a bit of trial and error to find how many Whole Plots are necessary to achieve 80% power for the 0.25 unit difference. With 24 whole plots, 48 runs, we are almost guaranteed of detecting the 0.75 unit difference and have power of about 70% for the 0.50 unit difference, but are still only near 40% for the 0.25 unit difference. Doubling the size of the experiment to 48 whole plots brings the power for the 0.25 unit difference to about 70%. At 72 whole plots, simulated power reaches around 85%. Finally, at 60 whole plots, the rejection rate for the 1000 simulations is just under 80%. At this point it is up to the researchers to decide whether their experiment budget allows for so many chips, or whether they will have to proceed with a smaller experiment and understand they will be less likely to detect their difference of interest.

Figure 8.7: Multiple Comparisons Simulation Power Analysis Results

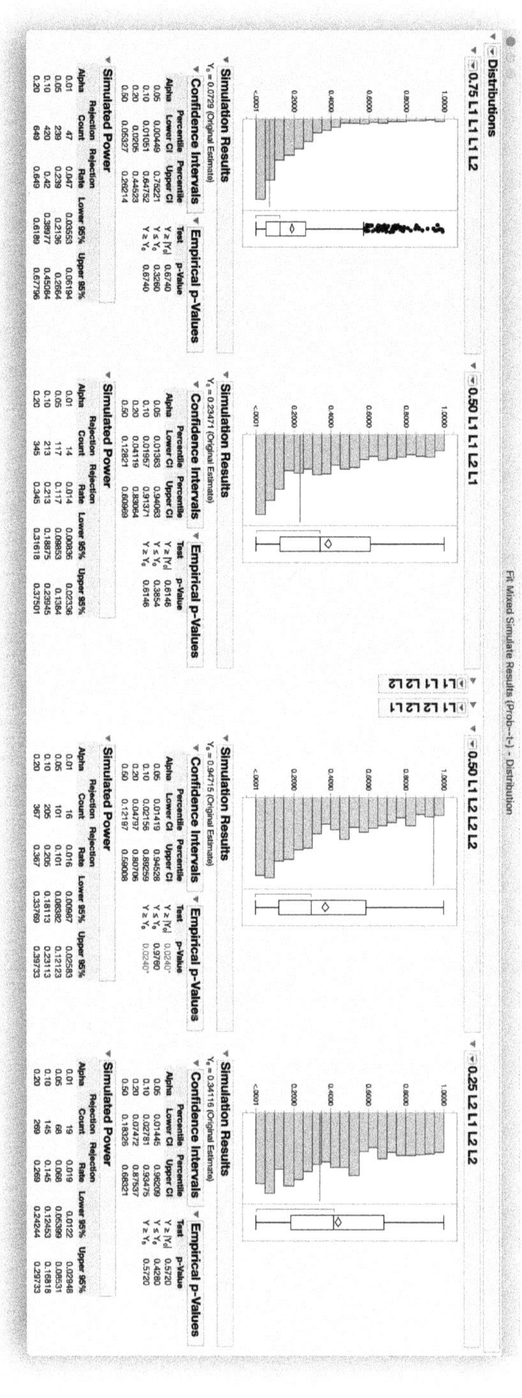

8.5 Confidence Interval Coverage Using the Winter Wheat Example

After following the creation of the simulation column in Section 8.2, you should have a `yield Simulation Formula` column in the `winterwheat.jmp` data table and a *Fit Mixed* report with the random coefficient model fit open. For this example, we are going to assume that the estimate for the effect of moisture, 0.6617, is the "true" effect of moisture and determine how often the confidence interval for moisture contained that true value. If you wanted to set a different value, you could edit the formula column and change it in the _Beta vector.

Determining confidence interval coverage requires a bit more work after completing the simulation. First, we have two simulation runs where we set a random seed the same for both runs. We obtain simulation results for the lower confidence bound and then the upper. We then combine those two results data tables into one and add a formula column to determine whether that interval contains the true value. Finally, run *Distribution* on the coverage formula column to see the percentage of intervals that contain the true value.

Confidence Interval Coverage Simulation Instructions

With the `winterwheat.jmp` random coefficient model report run, right-click on the *95% Lower* column in the Fixed Effects Parameter Estimates table of the report then choose *Simulate*.

A dialog box appears with the options for the simulation. The *Column to Switch Out* selects the column that was used as the Y, response, `yield` column, which is what we want for this simulation. The *Column to Switch In* selects the `yield Simulation Formula` column.

In the *Number of Samples* box change the number from the default 2500 to 1000. We then only have to count the number of successful coverages to determine coverage probability. *Distribution* does make it easy to calculate percentages, however, if you want to run a larger study.

Set the *Random Seed* to 8675309 to replicate these results, or your own random seed. Just be sure to note what you choose so that you will use the same one for the second run. (It will be saved in the results data table if you forget.)

Once the options are set, click *OK* to run the simulation.

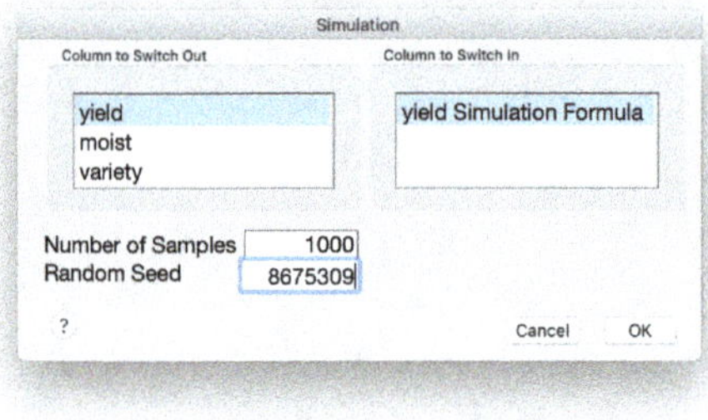

After the simulation is complete, repeat the same instructions with the *95% Upper* column. The reason we set the random seed the same in both runs is so that the data used for each sample run are the same. This ensures that the Lower and Upper confidence bounds were calculated from the same data, and we get the proper coverage. You should now have two Fit Mixed Simulate Results data tables, one for the lower and one for the upper confidence bound. We need to join these two tables together to have both bounds in the same table.

Joining Two Simulations Results Tables

In the `Fit Mixed Simulate Results (95% Lower)` data table window, choose *Tables > Join*. A Join dialog box will appear with the options for how to complete the join.

Choose `Fit Mixed Simulate Results (95% Upper)` in the *Join 'Fit Mixed Simulate Results (95% Lower)' with* box.

The `SimID` is common to both tables, so those are the columns that we want to match by. In the *Source Columns* section, select `SimID` in both data table columns lists and then click *Match* in the *Matching Specification* box.

We do not need all of the columns from the simulation results data tables, so click the check box for *Select columns for joined table* in the *Output Columns* box.

Either one at a time or holding down the Cntl key and clicking to select all at once, add `SimID` from one of the *Source Columns* lists and `moist` from both lists to the *Output Columns* by clicking *Select*.

If you want to preserve this dialog box after joining, click the check box for *Keep dialog open*. If the join does not perform as you wish (perhaps you accidentally omitted an output column), this saves what you did to try changing other options.

You can give the resulting table a name in the *Output table name:* text box, or leave it blank, and JMP will assign an Untitled name to it.

Once the options are set, click *OK* to join the tables.

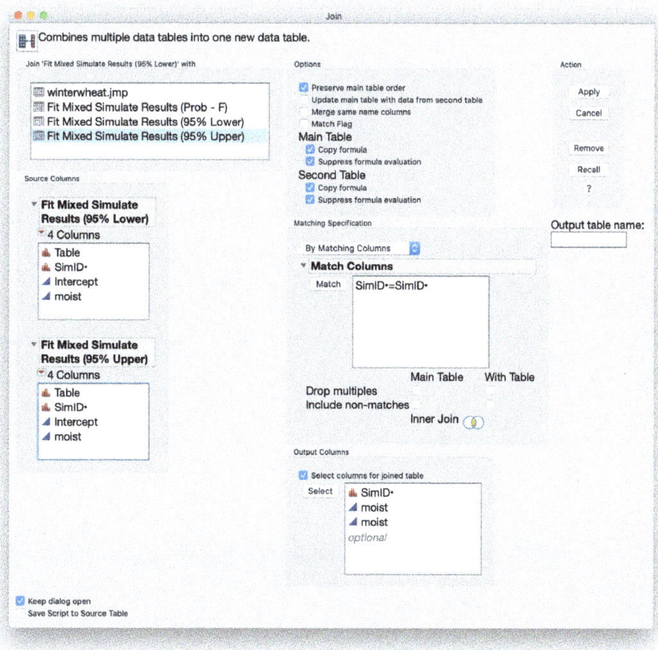

Figure 8.8: Joined Simulation Confidence Intervals Data Table

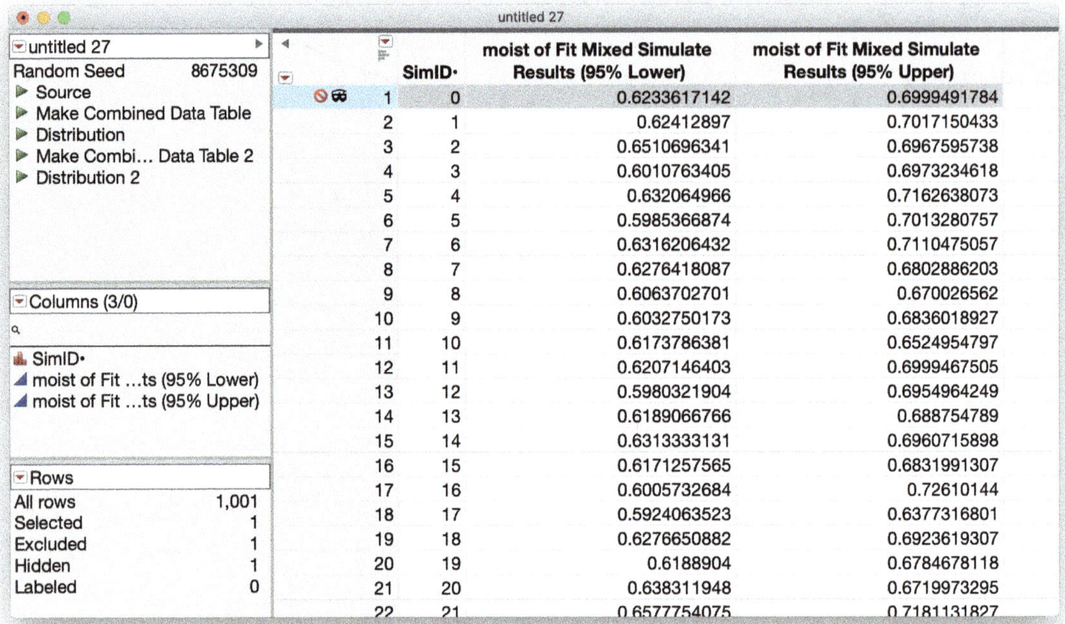

untitled 27

untitled 27
Random Seed 8675309
Source
Make Combined Data Table
Distribution
Make Combi... Data Table 2
Distribution 2

Columns (3/0)
SimID•
moist of Fit ...ts (95% Lower)
moist of Fit ...ts (95% Upper)

Rows
All rows 1,001
Selected 1
Excluded 1
Hidden 1
Labeled 0

	SimID•	moist of Fit Mixed Simulate Results (95% Lower)	moist of Fit Mixed Simulate Results (95% Upper)
1	0	0.6233617142	0.6999491784
2	1	0.62412897	0.7017150433
3	2	0.6510696341	0.6967595738
4	3	0.6010763405	0.6973234618
5	4	0.632064966	0.7162638073
6	5	0.5985366874	0.7013280757
7	6	0.6316206432	0.7110475057
8	7	0.6276418087	0.6802886203
9	8	0.6063702701	0.670026562
10	9	0.6032750173	0.6836018927
11	10	0.6173786381	0.6524954797
12	11	0.6207146403	0.6999467505
13	12	0.5980321904	0.6954964249
14	13	0.6189066766	0.688754789
15	14	0.6313333131	0.6960715898
16	15	0.6171257565	0.6831991307
17	16	0.6005732684	0.72610144
18	17	0.5924063523	0.6377316801
19	18	0.6276650882	0.6923619307
20	19	0.6188904	0.6784678118
21	20	0.638311948	0.6719973295
22	21	0.6577754075	0.7181131827

Figure 8.8 shows the table with the upper and lower confidence bounds from each of the simulations. Notice in this joined table, the first row with SimID 0 no longer has the hide and exclude row state. We do not want to include it in our Distribution, so right-click on row 1 and select *Hide and Exclude*. We now need to add a final column to the data table that calculates whether the "true value" of moisture is contained in the intervals. We do this by adding a formula.

Creating the Comparison Formula

In the joined data table window, choose *Cols* > *New Columns...*. A New Column dialog box will appear with the options for this new column.

Name the column what you want. We chose "Contains 0.661655446329778", as that is the value for the effect of moisture used in the simulation formula.

The *Data Type* is correct with the Numeric default, but change the *Modeling Type* to Nominal because the formula that we will create will return a 0, if the interval does not contain the true value, or 1, if it does.

Click *Column Properties* then *Formula* to add the formula. A formula dialog box will pop up to build the formula.

In the Columns list, double-click or drag `moist of Fit Mixed Simulate Results (95% Lower)` to add it to the formula. The name will probably be abbreviated, but you should be able to see the "moist" and "Lower" parts of the name.

The leftmost panel of the dialog box contains a sorted list of commonly used functions. Expand the *Comparison* section, and then choose *a* <= *b*. We want to compare whether the lower bound is less than or equal to the true value.

In the empty box to the right of the <= sign in the formula, type or paste the true value, 0.661655446329778.

Now click the *a* <= *b* comparison operation again. The second part of our comparison is whether the true value is less than or equal to the upper bound.

Finally, double-click or drag `moist of Fit Mixed Simulate Results (95% Upper)` and add the column to the formula box to the right of the second <= sign.

This completes the formula creation, shown below. Click *OK* to close the Formula dialog box and return to the New Column dialog box. With the formula complete and the options set, click *OK* to add the new column to the data table.

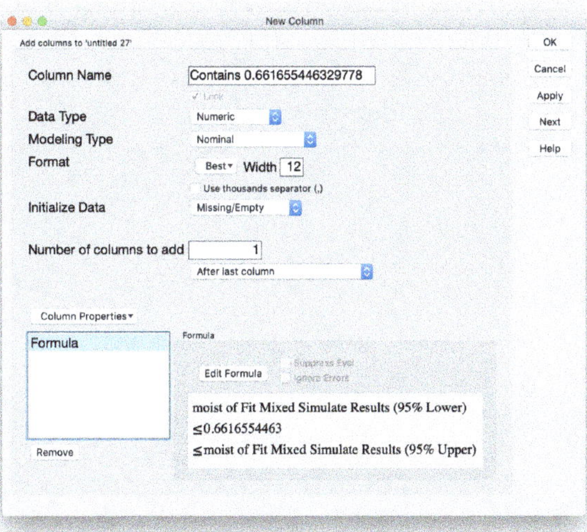

The formula column in the data table now contains ones and zeros reflecting whether or not the true value falls between the lower and upper confidence bounds. If we run *Distribution* on this column, we will know what percentage of those intervals contained the true value. This percentage is an empirical measure of confidence level.

Calculating Confidence Interval Coverage

In the joined data table window, choose *Analyze > Distribution* to launch the Distribution platform dialog box.
Add the **`Contains 0.661655446329778`** (or whatever you named it) to the *Y, Columns* role.
Click *OK* run the Distribution report.

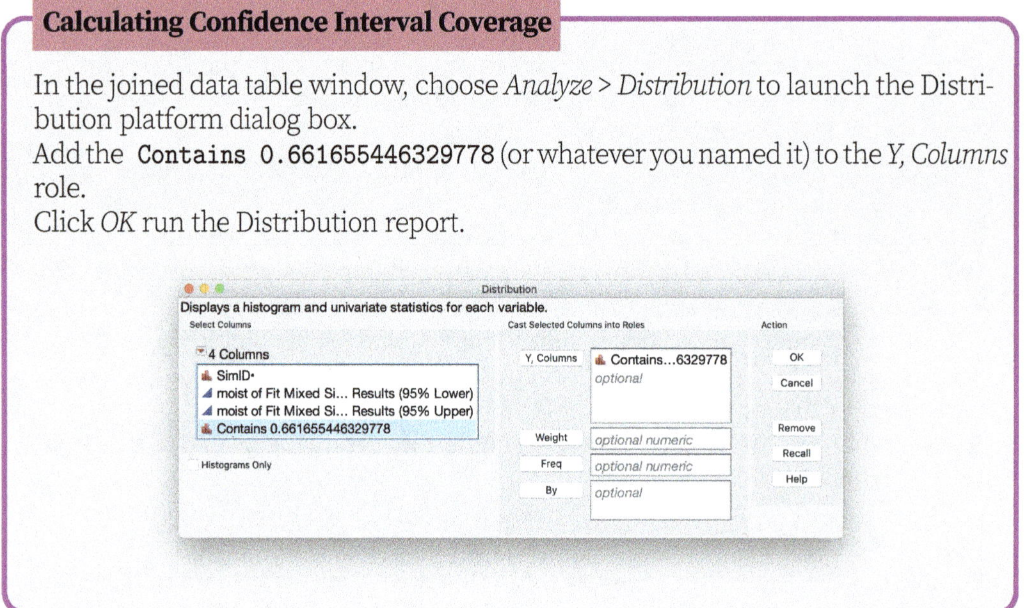

Figure 8.9 shows the Distribution report. We see that of the 1000 simulations, 1 had missing confidence interval information and 999 had intervals. Of the 999, 913 contained the true value, or 91.391%. This is slightly below the stated confidence level of 95% even accounting for sampling. This could indicate that the model is not quite correct. If you recall from Chapter 5, we determined that the covariance between the intercept and slope was negligible and should be omitted. That could also be the case with these data. We used a model that included the covariance, but that covariance may have been negligible. We leave it as an exercise to investigate this behavior further.

Figure 8.9: Confidence Interval Coverage Results

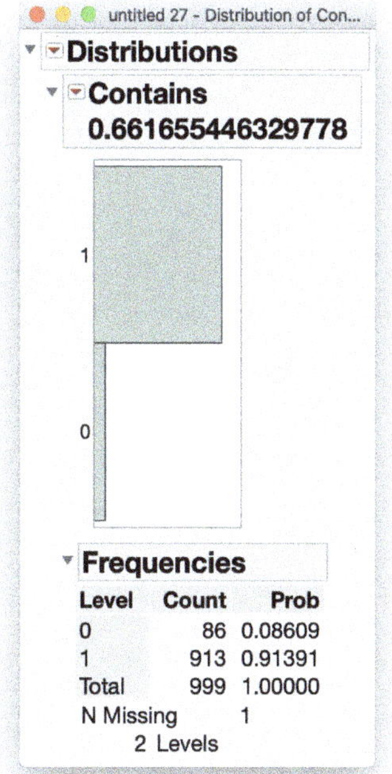

8.6 Simulating Mixed Model Data Directly with JSL

In addition to their primarily mouse-driven interfaces and interactive graphics, JMP and JMP Pro come prepacked with a rich, interpreted coding language known as JMP Scripting Language (JSL). In fact, the formulas that we have been creating and editing in previous sections utilize JSL functions. JSL has some similarities to the SAS language, with both originating at SAS Institute, Inc., being semicolon delimited, and sharing numerous function names. JSL's main purpose is to provide a way to script nearly everything doable with mouse clicks in the graphical user interface. Most red triangle menus contain an option near the bottom to Save Script in various ways, and this is an excellent way to begin learning JSL. Also, the Log in JMP 16 now displays JSL derived from most mouse clicks. To edit and run JSL scripts, JMP has a script editor with helpful features such as syntax highlighting, bracket matching, and mouse-over function tips. To open this editor, click the New Script icon near the upper left corner of the main JMP window.

We now illustrate how you can simulate and analyze mixed model data from scratch directly using JSL. This opens endless possibilities for various simulation scenarios of interest across scientific and engineering domains. The goal here is to provide a good starting block of code that you can subsequently modify to suit your needs. We simulate data from a basic Randomized Complete Block Design and then analyze it using Fit Model.

The setup of the simulation is as follows. Each RCBD data set has 10 blocks and 3 treatments for a total of 30 observations. The treatment effects are all set to 0 in a vector, so we are simulating under the full null hypothesis. The block variance component is set equal to 2 and the residual error variance to 1, and both sources of variability are assumed to be normally distributed as usual. We create 1000 simulations of this setup, and plan to examine the sampling distributions of the common mixed model statistics that we have been studying throughout this book.

Here is the JSL code, which is also available in the file **rcbd_sim.jsl**. The main sections of the code begin with comments in lines beginning with the prefix //.

```
// simulate data from rcbd

// inputs
trt = [0,0,0];
nb = 10;
ns = 1000;
s2b = 2;
s2e = 1;
seed = 123;

// derived variables
nt = length(trt);
sb = sqrt(s2b);
se = sqrt(s2e);
RandomReset(seed);

// create and populate a matrix
// columns are sim, block, trt, y
m = Matrix(ns*nb*nt,4);
i = 1;
for(s=1, s<=ns, s++,
   for(b=1, b<=nb, b++,
      u = sb*RandomNormal();
      for(t=1, t<=nt, t++,
         m[i,1] = s;
         m[i,2] = b;
         m[i,3] = t;
         m[i,4] = trt[t] + u + se*RandomNormal();
```

```
            i++;
        );
    );
);

// create jmp data table from matrix
d = AsTable(m, << ColumnNames({"sim","block","trt","y"}));
Column(d,"block") << SetModelingType("nominal");
Column(d,"trt") << SetModelingType("nominal");
d << SetName("rcbd_sim");

// fit mixed model to each simulated set
fm = Fit Model(
        By( :sim ),
        Y( :y ),
        Effects( :trt ),
        Random Effects( :block ),
        NoBounds( 1 ),
        Run(
             :y << {Summary of Fit( 1 ), Analysis of Variance( 0 ),
             Parameter Estimates( 1 ), Scaled Estimates( 0 ),
             Plot Actual by Predicted( 0 ), Plot Regression( 0 ),
             Plot Residual by Predicted( 0 ),
             Plot Studentized Residuals( 0 ),
             Plot Effect Leverage( 0 ),
             Plot Residual by Normal Quantiles( 0 ),
             {:trt << {LSMeans Student's t( 0.05,
                         Crosstab Report( 0 ),
                         Connecting Letters Report( 0 ),
                         Detailed Comparisons( 1 )
             )}}}
        ),
);

// extract results from reports
vce = Report(fm[1])["REML Variance Component Estimates"][Table Box( 1 )]
            << Make Combined Data Table;
fet = Report(fm[1])["Fixed Effect Tests"][Table Box( 1 )]
            << Make Combined Data Table;
lmd = Report(fm[1])["Effect Details"][Table Box( 2 )]
            << Make Combined Data Table;
```

```
vce << SetName("variance_component_estimates");
fet << SetName("fixed_effect_tests");
lmd << SetName("lsmean_diffs");

// add column identifying which difference
jsl = substitute(
          expr(
               lmd << NewColumn("diff", Numeric,
                    Formula(floor(mod(row()-1,_nk5_)/5)+1),
               )
          ),
          expr(_nk5_),
          5*NChooseK(nt,2)
);
jsl;


// distributions of fit model statistics

// variance component estimates
vce << Distribution(
          Continuous Distribution( Column( :Var Component ) ),
          By(:Random Effect)
);

// f-statistic p-values
fet << Distribution( Continuous Distribution( Column( :"Prob > F"n ) ) );

// lsmean diffs and confidence limits
lmd << Distribution(
           Continuous Distribution( Column( :Column 2 ) ),
           By(:diff,:Column 1)
);

// lsmean diff one- and two-sided p-values
lmd << Distribution(
          Continuous Distribution( Column( :Column 6 ) ),
          By(:diff,:Column 4)
);
```

The main steps in the JSL code are:

1. Define the RCBD structure with editable input variables.
2. Simulate data in a large JMP matrix.
3. Convert the data to a JMP data table.
4. Analyze each simulated set with `Fit Model()` using `By()`.
5. Assemble statistics of interest from each report into new JMP tables.
6. Examine distributions of the statistics to assess the simulation results.

A few keys to the JSL code that may not be obvious at first glance are as follows:

- The loop generating the block random effects is outside that of the treatment effects so that the block random effect can be held constant for all observations in that block.
- The simulated data are created at first in a JMP matrix, and this is then used to populate a JMP table. JSL has a full collection of matrix routines that can be invaluable when simulating more complex mixed models.
- The `Fit Model()` call contains specific commands to create reports that we want, for example **`Detailed Comparisons`** of treatment least squares means. The exact syntax for such calls can be obtained easily by first running a desired analysis with the mouse on a single data set, and then clicking the red triangle > *Save Script* > *To Script Window* or by examining the JMP Log (Ctrl-L).
- `« Make Combined Data Table` is a convenient way to extract results from all reports at once. Reports can be referenced by the title in the outline box containing them.
- The LSMeans details output table contains a complete set of difference statistics and is not quite in the form that we need to do proper distributions. The code adds a new column named **`diff`** to address this, utilizing the `mod()` and `floor()` functions.
- The JMP Distribution platform is a great way to initially explore simulation results. We use the `By()` command within a `Distribution()` call whenever statistics are stacked into a single column with one or more other columns identifying the unique statistic.

To run this code, open the file **`rcbd_sim.jsl`** in the JMP script editor, and click the green Run Script arrow. The code takes 5-10 seconds to execute and generates several windows. We now consider the outputs in the order they are created. Use the JMP Home Window (Ctrl-1) or native features on your computer to navigate among the windows.

Figure 8.10 shows the simulated RCBD data. We have 1000 simulations, 10 blocks, and 3 treatments, stacked into a total of 30000 observations.

Figure 8.10: Simulated RCBD Data

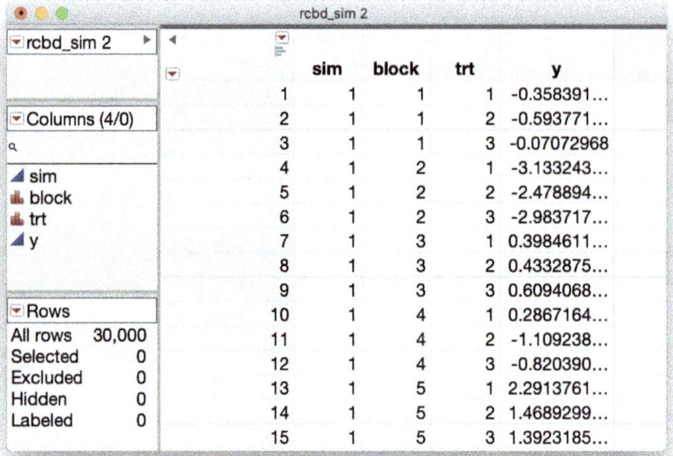

rcbd_sim 2

rcbd_sim 2

Columns (4/0)

sim
block
trt
y

Rows

All rows	30,000
Selected	0
Excluded	0
Hidden	0
Labeled	0

	sim	block	trt	y
1	1	1	1	-0.358391...
2	1	1	2	-0.593771...
3	1	1	3	-0.07072968
4	1	2	1	-3.133243...
5	1	2	2	-2.478894...
6	1	2	3	-2.983717...
7	1	3	1	0.3984611...
8	1	3	2	0.4332875...
9	1	3	3	0.6094068...
10	1	4	1	0.2867164...
11	1	4	2	-1.109238...
12	1	4	3	-0.820390...
13	1	5	1	2.2913761...
14	1	5	2	1.4689299...
15	1	5	3	1.3923185...

Figure 8.11 shows the main output from `Fit Model()`. This window contains 1000 reports and is suitable only for spot checking a few individual simulation results and for seeing how the output is named and arranged within each report.

Figure 8.11: Fit Model Output from Simulated RCBD Data

rcbd_sim 2 - Fit Least Squares

Response y sim=1

Effect Summary

Summary of Fit

RSquare	0.881406
RSquare Adj	0.872622
Root Mean Square Error	0.575315
Mean of Response	0.175676
Observations (or Sum Wgts)	30

Parameter Estimates

Random Effect Predictions

REML Variance Component Estimates

Random Effect	Var Ratio	Var Component	Std Error	95% Lower	95% Upper	Wald p-Value	Pct of Total
block	4.6617089	1.5429657	0.7802378	0.0137277	3.0722037	0.0480*	82.337
Residual		0.3309871	0.110329	0.1889773	0.7238431		17.663
Total		1.8739528	0.7828337	0.9513507	5.2582519		100.000

-2 LogLikelihood = 80.237938176

Note: Total is the sum of the positive variance components.

Total including negative estimates = 1.8739528

Covariance Matrix of Variance Component Estimates

Iterations

Fixed Effect Tests

Source	Nparm	DF	DFDen	F Ratio	Prob > F
trt	2	2	18	1.9884	0.1659

Effect Details

Response y sim=2

Effect Summary

Summary of Fit

RSquare	0.808031

Figure 8.12 shows the distribution of the 1000 block variance component estimates from the simulated data. In this and subsequent results, your results may vary slightly from those shown here due to random number generator differences. The mean value is very close to the true value of 2, as expected, and the overall distribution is skewed toward larger values as expected from a mixture of gamma distributions. Also note a small proporition of negative values, which are permitted as specified by the `NoBounds(1)` option in the JSL code. Allowing negative values like this ensures that the variance component estimates are unbiased and that fixed-effect tests and intervals retain their nominal rates.

Figure 8.12: Distribution of Block Variance Component Estimates for Simulated RCBD Data

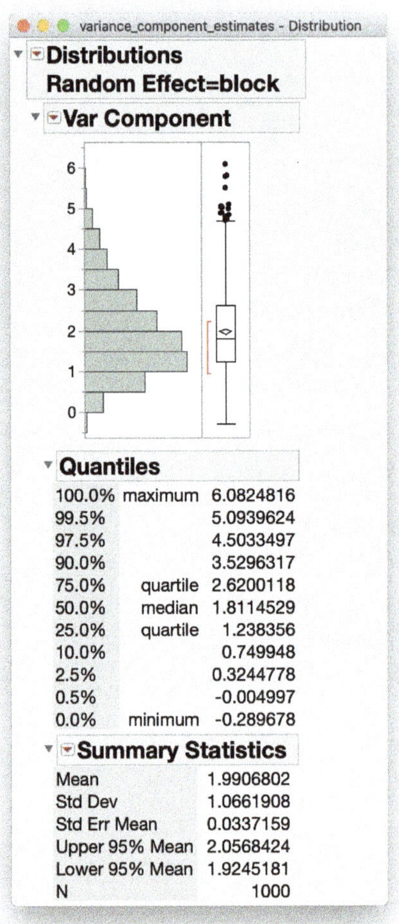

Figure 8.13 shows the distribution of the overall treatment F-statistic p-values. Under the null hypothesis of no treatment effect, we expect this distribution of p-values to be uniformly distributed, and the observed distribution agrees to a reasonable degree of approximation. Along with this, the fraction of p-values falling below any specific cutoff like 0.05 will be approximately the same fraction, demonstrating that the F test is holding its nominal size. This result provides a basis for performing valid power calculations in which the true treatment effects are set to nonzero values.

Figure 8.13: Distribution of F test p-values for Simulated RCBD Data

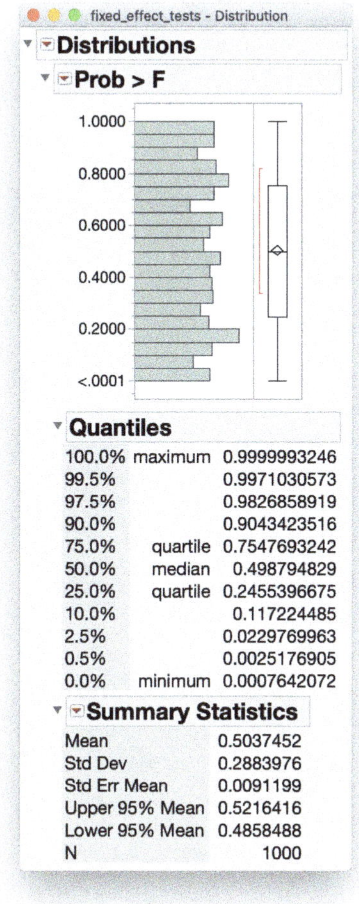

Figure 8.14 shows the distribution of the simulated first LSMean difference. As expected, the difference has a Student t sampling distribution with a mean of 0.

Figure 8.14: Distribution of First LSMean Difference for Simulated RCBD Data

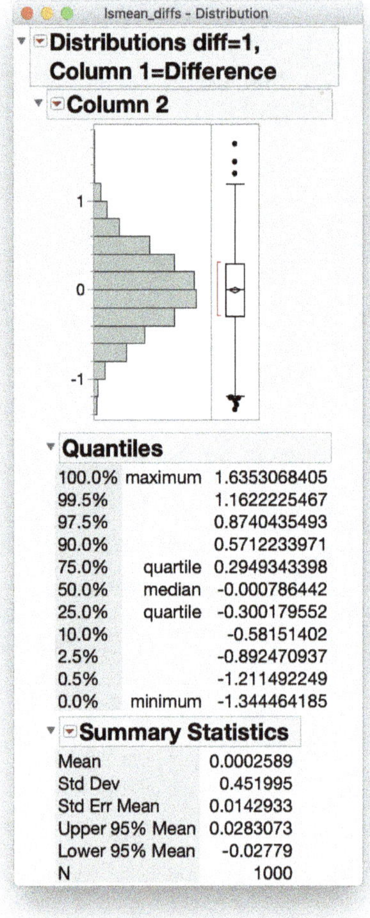

Additional results (not shown) further verify that the full RCBD simulation is performing as expected. As mentioned above, this code and analysis is intended only to be a starting point, illustrating one way to perform simulations directly with JSL. This specific simulation demonstrates that mixed model statistics are performing as claimed under standard assumptions, with means of estimates displaying expected true values and p-values verifying nominal test size under the null hypothesis of no treatment effect. In general, simulations are an invaluable way to explore deeper questions that you may have about how mixed models are performing in specific scenarios. You can quantify how robust various output statistics are to various types of model misspecification you may encounter in practice. Refer to Gibbs and Kiernan (2020) for specific ways to simulate from more complex mixed models and Wicklin (2013) for a great introduction to simulation in general.

8.7 Exercises

1. If the researchers in the semiconductor experiment absolutely could not use more than 48 whole plots, what might be another way for them to increase the power of the test for the 0.25 unit difference?
2. Using the `winterwheat.jmp` data table and a simulation formula that includes the covariance term, determine the confidence interval coverage for the moisture coefficient when analyzing the data with a model that does not include the covariance. Hint: you will need to save the simulation formula from a full model and run Fit Mixed again using a reduced model before running the simulation study.
3. Using the `winterwheat.jmp` data table and a simulation formula that does not include the covariance term, determine the confidence interval coverage coefficient when analyzing the data with that same reduced model. How does this coverage compare to the coverage in the previous exercise? Why do you think there's a difference (if there is)?
4. In the JSL RCBD simulation, what is the true expected value of the standard deviation of the LSMean differences? How do the observed simulation values compare?
5. In the JSL RCBD simulation, examine the sampling distributions of the 95% lower and upper confidence limits of LSMean differences. What fraction of values do we expect to show opposite sign in each case? How do the observed simulation values align with expected ones?
6. In the JSL RCBD simulation code, modify the treatment effect vector, `trt=[0,0,0]`, to contain nonzero values and rerun the entire simulation. How do results change? Provide power estimates for two different tests.
7. Using a formula column or JSL, simulate data like the cell viability data from Chapter 1 using mixed model assumptions and parameters from the original analysis. Analyze the simulation distributions and compare results to the original analysis.

Chapter 9

Generalized Linear Mixed Models

Throughout this book we have thus far considered examples for which the response is continuous, assuming that the random effects and residual errors follow a normal (Gaussian) distribution. While this is arguably the most common type of mixed model, in many cases the response is discrete. In this chapter, we address mixed models for non-normally distributed data.

9.1 Motivating Examples

The three most common examples of non-normal data are:

Binomial — the number of successes, y, out of n trials. The **Shrub Coverage** example from Parks Canada measures the number of sampling points out of 100 that contain species from a plant functional group during a given year. A "success" in this case is the presence of the functional group, and a "trial" is the examination of a plot for that type of plant.

Binary — (also known as *Bernouilli*), a special case of binomial with only two distinct values such as [0,1], [Yes,No], or [True,False]. The **Salamander Mating** example is of this type. For these data, the response is an indicator of whether mating occurred in that experimental group. Related experimental features are the day on which the experiment occurred, the population group from which the females were taken (of two groups), the population group from which the males were taken (of the same two groups), and the numbers of females and males present in the experimental group.

Count — with integer values 0,1,2,.... The **Manufacturing Imperfections** example represents a common type of data from industrial quality control. Here the counts are the number of defects in a manufactured part, and treatments represent different modes of production.

9.2 Conceptual Background

How are we to fit mixed models to such data? One basic approach that works well in many cases is to simply transform the response and assume that it follows a standard normal linear mixed model. For example, many laboratory assay measurements are positive, have a distribution skewed to the right, and major sources of variability can reasonably be assumed to be multiplicative. In this case, taking the log of the response and fitting mixed models as in previous chapters is perfectly sensible and straightforward. When taking this approach, choose the transform with care and realize it impacts all components of the mixed model, including unit-level residual errors.

To put non-normal data and associated transformations on a solid theoretical foundation, the elegant statistical theory described in the classic book by McCullagh and Nelder (1989) encompasses the most common probability distributions used in statistical modeling in an *exponential family*. The theory is called *generalized linear models* and includes distributions such as Gaussian, Bernouilli, Binomial, and Poisson. The techniques of generalized linear models describe a unified approach to modeling such data with link and deviance functions as well as practical ways for performing statistical inference.

In the early 1990s, the theory was extended to mixed models independently by various researchers, including Schall (1991), Breslow and Clayton (1993), and Wolfinger and O'Connell (1993). A natural name for the extension is *generalized linear mixed models* (GLMMs). The research led quickly to GLMM software: the SAS GLIMMIX macro, PROC GLIMMIX in SAS/STAT, and routines in several other statistical packages. Stroup et al. (2018), Stroup (2012), and Bolker et al. (2009) provide a rich and detailed collection of practical examples and analytical wisdom for handling GLMMs using PROC GLIMMIX.

GLMMs provide a great way to handle the common phenomenon of *overdispersion*. This is when non-normal data exhibit variability exceeding that expected from the assumed probability distribution with independent observations. Random effects introduce a formal mechanism for modeling correlations within clusters of observations, often completely explaining overdispersion evident when random effects are ignored and observations are naively assumed to be independent.

An intuitive way to think about a GLMM response is as a discretized transform of a normal linear mixed model response. For example, begin with a normal linear mixed model from some designed experiment. Apply a logistic transform to y, which puts it on a 0 to 1 scale, and then apply a threshold to make the response itself either 0 or 1. We now have a binary response mixed model. The transform turns out to be the inverse of the *link function* in GLMM parlance, which in this example is the logit function.

The three main elements of a GLMM are:

1. **Distribution**: The assumed probability distribution of the response, for example, binomial or Poisson.
2. **Linear Predictor**: A linear mixed model expression that accommodates the experimental setup. For example, $\eta_{ij} = \mu + t_i + b_j$ would be an appropriate linear predictor for a randomized complete block design.

3. **Link Function**: An invertible function that connects the linear predictor to the expected value of the assumed distribution, conditional on the random effects.

The link function is the central connector of a GLMM. The theory of generalized linear models describes *canonical* link functions that are the most mathematically convenient and understandable ones for deriving results. For the binomial distribution, the canonical link function is the logit, and its inverse is the logistic function. For the Poisson distribution, the canonical link function is the log, and its inverse is the exponential function. Other members of the exponential family have corresponding canonical link functions that map linear predictors to the scale of the observed response. Although it is possible to use link functions other than the canonical ones, these tend to be for more advanced, complex situations beyond the scope of the introductory examples covered herein.

Whenever the link function is nonlinear like logit or log, GLMMs become a special type of *nonlinear mixed model*. Nonlinear mixed models are not covered generally in this book, but have a strong theoretical background and history of applications (e.g., Sall (2014)). The nonlinearity of GLMM link functions poses new specific challenges for fitting and interpretation, and several different ways exist for carrying out the computational and statistical analysis.

One common way of fitting GLMMs builds on the aforementioned transformation intuition and is known as *residual pseudo likelihood* (RSPL). The idea is to use a Taylor series linear expansion of the inverse link function to create a pseudo-response and weights. This pseudo-response can subsequently be treated as the Y variable in JMP in the usual mixed model, as if that weighted pseudo-response were normally distributed. The pseudo-response and weights are then updated after the first fit, and the whole process is repeated until convergence. This iterative algorithm works well in most common cases, but is approximate and can produce biased results in some situations. The worst bias tends to be in cases that are radically non-normal, for example, binary repeated measures with only a few measurements per subject (Stroup and Claassen, 2020).

While GLMMs are not yet officially implemented in JMP or JMP Pro, a free add-in is available in Dong (2020) that implements RSPL. The final output from the add-in is the last iteration from the mixed model fit on the normally distributed weighted psuedo-response, and is on the scale of the linear predictor. Interpretation of results is similar to that in previous chapters, but keep in mind the method is using an approximate response on a transformed scale. We now explore three examples of the add-in that show how you can use it in practice.

9.3 Binomial Response: Shrub Coverage

This example uses the table `shrub.jmp`, which contains binomial Plant Community Monitoring Data from the Auyuittuq National Park in Nunavut, collected by Parks Canada (Samson, 2021). A special thanks to Ecological Modeling Specialist Claude Samson for providing the data and as well as statistical guidance. Fifteen different plots are examined in four different years for the presence/absence of three functional groups (shrubs, lichens, graminoids), producing binominal responses out of 100 samples. The binomial counts are plotted in Figure 9.1.

Figure 9.1: Line Plots of Shrub Coverage Data

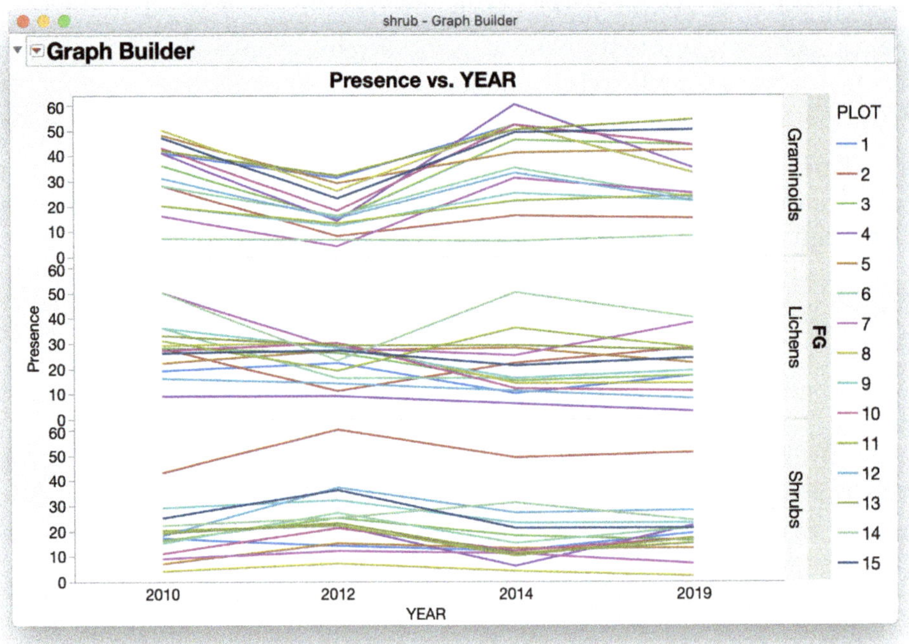

A primary research question is if the functional group coverages have changed significantly across the three years, and if so, how. The line plots in Figure 9.1 reveal an increase in graminoids from 2012 to 2014 and a slight downward trend in lichens. The plots exhibit a variety of patterns along with a good degree of consistency within a plot. This is a block design with 3 x 4 factorial fixed effects `FG`, `YEAR`, and `FG*YEAR`. For random effects, we specify `PLOT` and `FG*PLOT`. The random effect `PLOT` accounts for a common covariance among all measurements from the same plot, and the random effect `FG*PLOT` accounts for an additional common covariance among observations from the same functional group and plot. Based on the lines graph, we might expect the random effect for `FG*PLOT` to exhibit a fairly strong variance component and the fixed effect for `FG*YEAR` to be significant given the different patterns over time for the three functional groups.

GLMM for the Shrub Coverage Data

$$y_{ijk}|(b_k,g_{ik}) \sim \text{Binomial}(n_{ijk},\pi_{ijk})$$
$$\eta_{ijk} = \mu+\phi_i+\tau_j+\phi\tau_{ij}+b_k+g_{ik} = \text{Logit}(\pi_{ijk})$$

y_{ijk} is the ijk^{th} observation, which has a binomial distribution conditional on the random effects.
n_{ijk} is the number of trials for y_{ijk}, in this case 100 for all i,j,k.
π_{ijk} is the probability of presence for the ith functional group, j^{th} year, and k^{th} plot.
η_{ijk} is the linear predictor.
μ is the overall intercept.
ϕ_i is the fixed effect of the ith functional group.
τ_j is the fixed effect of the jth year.
$\phi\tau_{ij}$ is the interaction fixed effect of the ith functional group and j^{th} year.
b_k is the random effect of the kth plot, $b_k \sim N(0,\sigma_b^2)$.
g_{ik} is the random effect of the ith functional group and k^{th} plot, $g_{ik} \sim N(0,\sigma_g^2)$.

Note the binomial distribution has no specific residual error, but the two random effects account for overdispersion.

Running the GLMM Add-In on the Shrub Data

To fit a GLMM to the shrub data, install the Generalized Linear Mixed Model add-in from Dong (2020), click *Add-Ins > Generalized Linear Mixed Model*, and set up the add-in dialog as below.

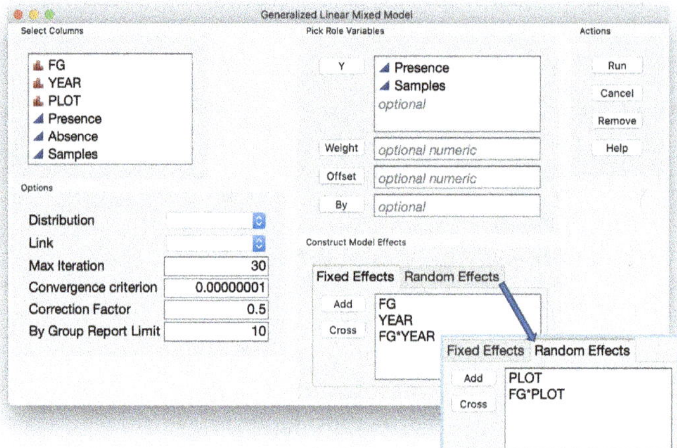

For a binomial response, we have two different variables: the number of successes and the number of trials. The add-in has a convention that you specify both variables the *Y* field, with successes first, then number of trials. In this case `Presence` is the number of successes and `Samples` is the number of trials. The fixed effects are `FG`, `YEAR`, and `FG*YEAR`. Specify the latter by selecting both `FG` and `YEAR` on the left side and then clicking the *Cross* button. Note `YEAR` is nominal to accomodate any possible pattern over the four years. The random effects are `PLOT` and `FG*PLOT`; specify them on the *Random Effects* tab.

Click *Run* to fit the GLMM and obtain results as in Figure 9.2.

As anticipated from the line plots, the variance component estimate for `FG*PLOT` is substantial, almost thirty times larger than the residual variance estimate. Interestingly, the variance component estimate for `PLOT` is negative, providing some evidence of competition for resources among the three functional groups.

The interaction fixed effect `FG*YEAR` is strongly significant. Whenever a fixed effect interaction is large, the meaning of the corresponding main effects becomes somewhat muddled because they average over heterogeneous effects. Here we focus primarily on the interaction and conclude the functional groups are behaving differently over the years.

To further explore why the interaction effect is significant, it is very useful to make an

Figure 9.2: GLMM Add-in Results for Shrub Coverage Data

shrub - Fit Least Squares 2

Response _pseudo

Weight: _w

Effect Summary

Source	LogWorth	PValue
FG*YEAR	16.228	0.00000
YEAR	3.069	0.00085
FG	0.797	0.15953

Remove Add Edit FDR

Summary of Fit

RSquare	0.851896
RSquare Adj	0.842141
Root Mean Square Error	0.130833
Mean of Response	-1.02624
Observations (or Sum Wgts)	30.89923

Parameter Estimates

Random Effect Predictions

REML Variance Component Estimates

Random Effect	Var Ratio	Var Component	Std Error	95% Lower	95% Upper	Wald p-Value	Pct of Total
PLOT	-7.344982	-0.125725	0.049041	-0.221844	-0.029607	0.0104*	0.000
FG*PLOT	28.475934	0.4874272	0.1402185	0.212604	0.7622503	0.0005*	96.607
Residual		0.0171172	0.002191	0.0135204	0.022376		3.393
Total		0.5045443	0.1401097	0.3127458	0.9484386		100.000

-2 LogLikelihood = -70.48634869

Note: Total is the sum of the positive variance components.

Total including negative estimates = 0.3788191

Covariance Matrix of Variance Component Estimates

Iterations

Fixed Effect Tests

Source	Nparm	DF	DFDen	F Ratio	Prob > F
FG	2	2	27.06	1.9658	0.1595
YEAR	3	3	122.1	5.9017	0.0009*
FG*YEAR	6	6	122.1	21.2360	<.0001*

Effect Details

interaction plot of the least squares means. To do this, in the Effect Details section, click the red triangle next to `FG*YEAR` and select *LSMeans Plot*. Check the box to *Create an Interaction Plot* and choose `FG` as the overlay term. You should obtain a graph as in Figure 9.3.

We see a large increase in graminoids from 2012 to 2014 to be one key driver of the significant interaction. Overall, the three functional groups have quite different patterns

Figure 9.3: GLMM Add-in LSMeans Interaction Plot for Shrub Coverage Data

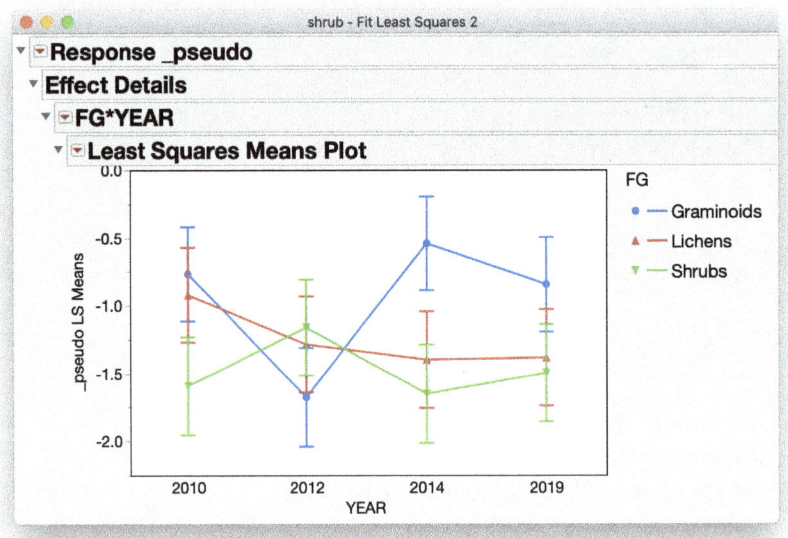

over the four years. Lichens exhibit a small but steady downward trend whereas shrubs exhibit alternate increases and decreases opposite of those from graminoids.

We consider `YEAR` to be nominal in this example in order to pick up any possible pattern across the different years. As more data are collected in future years, it could make sense to look and test for linear or more complex trends.

9.4 Binary Response: Salamander Mating

Our next example involves a binary response in a well-studied data set on salamander mating. Females and males sampled from two different populations of salamanders are measured on whether they mate. The data are available in `salamander.jmp` as shown in Figure 9.4.

The data have columns for the day of the experiment, female and male populations, female and male individual number, and three different binary responses. Here we analyze only the first response, y_1.

Figure 9.4: Salamander Mating Data

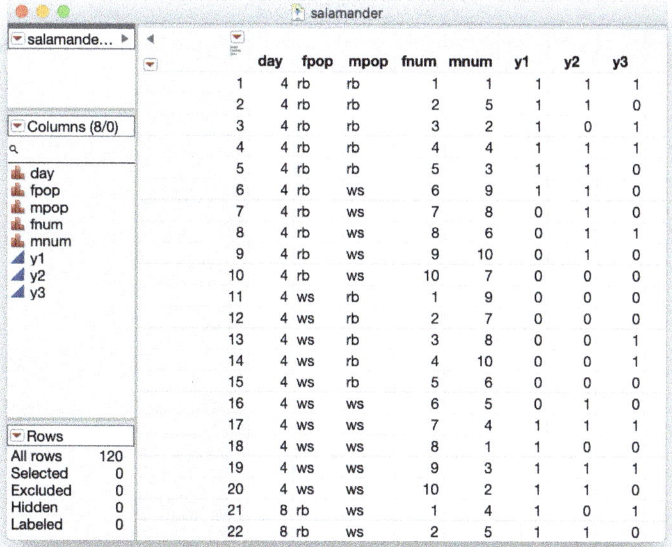

salamander

salamande...

Columns (8/0)

day
fpop
mpop
fnum
mnum
y1
y2
y3

Rows

All rows	120
Selected	0
Excluded	0
Hidden	0
Labeled	0

	day	fpop	mpop	fnum	mnum	y1	y2	y3
1	4	rb	rb	1	1	1	1	1
2	4	rb	rb	2	5	1	1	0
3	4	rb	rb	3	2	1	0	1
4	4	rb	rb	4	4	1	1	1
5	4	rb	rb	5	3	1	1	0
6	4	rb	ws	6	9	1	1	0
7	4	rb	ws	7	8	0	1	0
8	4	rb	ws	8	6	0	1	1
9	4	rb	ws	9	10	0	1	0
10	4	rb	ws	10	7	0	0	0
11	4	ws	rb	1	9	0	0	0
12	4	ws	rb	2	7	0	0	0
13	4	ws	rb	3	8	0	0	1
14	4	ws	rb	4	10	0	0	1
15	4	ws	rb	5	6	0	0	0
16	4	ws	ws	6	5	0	1	0
17	4	ws	ws	7	4	1	1	1
18	4	ws	ws	8	1	1	0	0
19	4	ws	ws	9	3	1	1	1
20	4	ws	ws	10	2	1	1	0
21	8	rb	ws	1	4	1	0	1
22	8	rb	ws	2	5	1	1	0

GLMM for the Salamander Data

$$y_{ijklm} | (d_k, f_{il}, m_{jm}) \sim \text{Bernouilli}(\pi_{ijklm})$$
$$\eta_{ijklm} = \mu + \alpha_i + \beta_j + \alpha\beta_{ij} + d_k + f_{il} + m_{jm} = \text{Logit}(\pi_{ijklm})$$

y_{ijklm} is the binary response of the $ijklm^{th}$ observation, which has a Bernouilli distribution conditional on the random effects.
π_{ijklm} is the probability of a favorable outcome for the $ijklm^{th}$ observation.
η_{ijklm} is the linear predictor.
μ is the overall intercept.
α_i is the fixed effect of the ith female population.
β_j is the fixed effect of the jth male population.
$\alpha\beta_{ij}$ is the interaction between the ith female population and j^{th} male population.
d_k is the random effect of the kth day, $d_k \sim N(0, \sigma_d^2)$.
f_{il} is the random effect of the lth female of the ith population, $f_{il} \sim N(0, \sigma_f^2)$.
m_{jm} is the random effect of the mth male of the jth population, $m_{jm} \sim N(0, \sigma_m^2)$.

The model has a 2×2 factorial structure for the fixed effects and three different random effects. Note the number coded for each individual female or male resets for each population, which is why we use a double subscript for f_{il} and m_{jm}.

Running the GLMM Add-In on the Salamander Data

To set up this analysis, click *Add-Ins > Generalized Linear Mixed Model,* and complete the dialog as below.

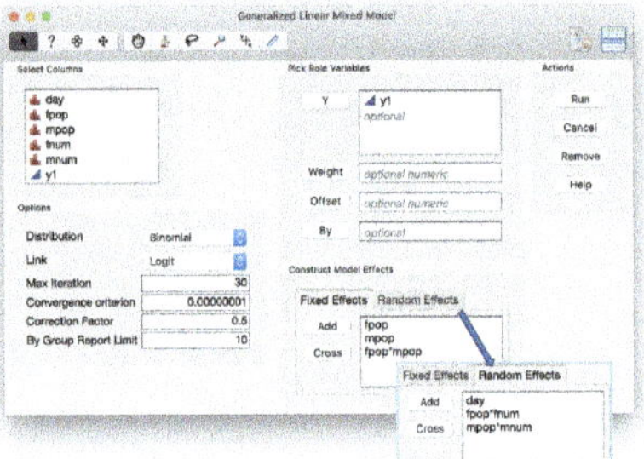

We specify **y_1** as the response and a 2×2 factorial for the fixed effects (**fpop**, **mpop**, and **fpop*mpop**). To specify the fixed effects in the dialog box, highlight **fpop** and **mpop** on the left, then click the *Add* button, then the *Cross* button. The random effects are **day**, **fpop*fnum**, and **mpop*mnum**. The latter two are set up to model the covariances due to the population structure in the data. Click the *Random Effects* tab and specify these three effects as shown above.

Click *Run* to fit the GLMM. As the add-in runs through each of its iterations, new columns in the data table are created and updated corresponding to the pseudo-response (named _pseudo), a weight variable (_w), and several other supplementary variables. See also the JMP log for an iteration history (not shown). The results are shown in Figure 9.5.

These results are from the final mixed model fit on the pseudo data and weights. Results are approximate but reasonable for drawing inferences from the experiment. The variance component estimates reveal relatively little day-to-day variability and approximately three times more variability in the females versus the males. This is an indication of higher selectivity among the females. As for the fixed effects, there is a relatively strong interaction between the two populations. Expand the Effect Details outline box to reveal and explore further details about the interaction.

Figure 9.5: GLMM Add-in Results for Salamander Mating Data

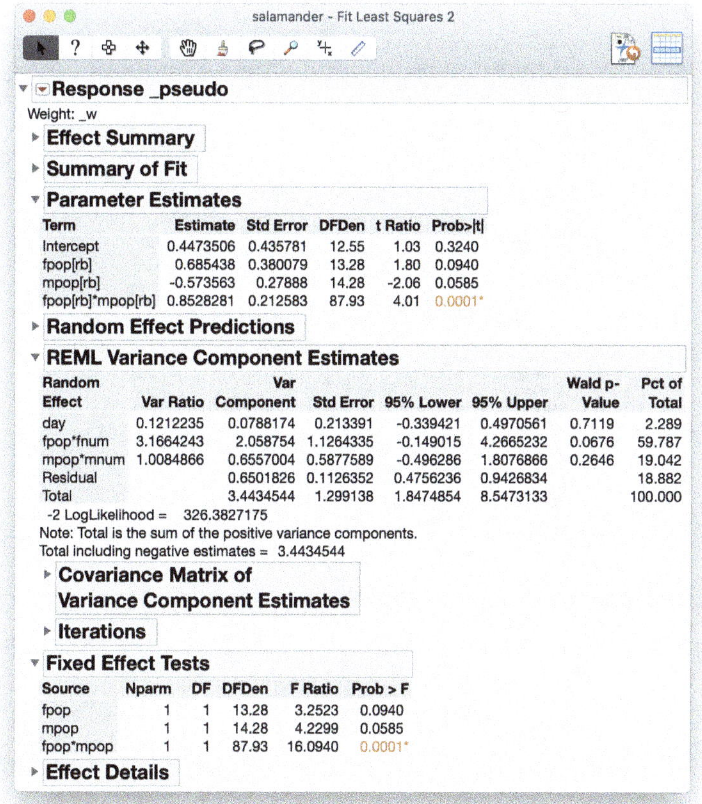

salamander - Fit Least Squares 2

Response _pseudo

Weight: _w

Effect Summary

Summary of Fit

Parameter Estimates

Term	Estimate	Std Error	DFDen	t Ratio	Prob>|t|
Intercept	0.4473506	0.435781	12.55	1.03	0.3240
fpop[rb]	0.685438	0.380079	13.28	1.80	0.0940
mpop[rb]	-0.573563	0.27888	14.28	-2.06	0.0585
fpop[rb]*mpop[rb]	0.8528281	0.212583	87.93	4.01	0.0001*

Random Effect Predictions

REML Variance Component Estimates

Random Effect	Var Ratio	Var Component	Std Error	95% Lower	95% Upper	Wald p-Value	Pct of Total
day	0.1212235	0.0788174	0.213391	-0.339421	0.4970561	0.7119	2.289
fpop*fnum	3.1664243	2.058754	1.1264335	-0.149015	4.2665232	0.0676	59.787
mpop*mnum	1.0084866	0.6557004	0.5877589	-0.496286	1.8076866	0.2646	19.042
Residual		0.6501826	0.1126352	0.4756236	0.9426834		18.882
Total		3.4434544	1.299138	1.8474854	8.5473133		100.000

-2 LogLikelihood = 326.3827175

Note: Total is the sum of the positive variance components.

Total including negative estimates = 3.4434544

Covariance Matrix of Variance Component Estimates

Iterations

Fixed Effect Tests

Source	Nparm	DF	DFDen	F Ratio	Prob > F
fpop	1	1	13.28	3.2523	0.0940
mpop	1	1	14.28	4.2299	0.0585
fpop*mpop	1	1	87.93	16.0940	0.0001*

Effect Details

9.5 Count Response: Manufacturing Imperfections

We now consider a count response in the data set `manufacturing_imperfections.jmp`. This a simulated industrial example of a balanced incomplete block design with 120 observations, 15 blocks, and 6 treatments. Each row represents a manufactured product, and the response Count is the number of defects found after a thorough inspection of that product. The first several rows of the data are shown in Figure 9.6.

Figure 9.6: Manufacturing Imperfections Data

Manufacturing Imperfections

Manufactur...
Design Balanced
Distribution
BIBD

Columns (3/0)
Block
Treatment
Count

Rows

All rows	120
Selected	0
Excluded	0
Hidden	0
Labeled	0

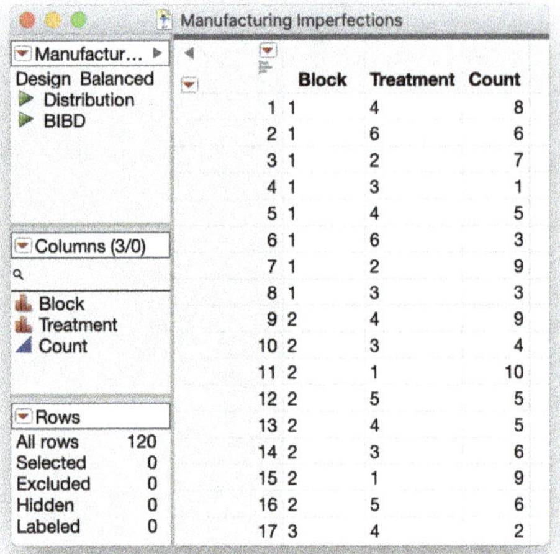

	Block	Treatment	Count
1	1	4	8
2	1	6	6
3	1	2	7
4	1	3	1
5	1	4	5
6	1	6	3
7	1	2	9
8	1	3	3
9	2	4	9
10	2	3	4
11	2	1	10
12	2	5	5
13	2	4	5
14	2	3	6
15	2	1	9
16	2	5	6
17	3	4	2

GLMM for the Manufacturing Imperfections Data

$$y_{ij}|b_j \sim \text{Poisson}(\lambda_{ij})$$
$$\eta_{ij} = \mu + \tau_i + b_j = \log(\lambda_{ij})$$

y_{ij} is the count response of the ij^{th} observation, which has a Poisson distribution conditional on the random effect.
λ_{ij} is the expected value of the ith treatment in the jth block.
η_{ij} is the linear predictor, which is on the log scale
μ is the overall intercept.
τ_i is the fixed effect of the ith treatment.
b_j is the random effect of the jth block, $b_j \sim N(0, \sigma_b^2)$.

As with the binomial and binary responses, there is no direct term for residual error, but the random effect provides a way to account for overdispersion due to blocking.

Running the GLMM Add-In on the Manufacturing Imperfections Data

Set up the GLMM add-in dialog as below.

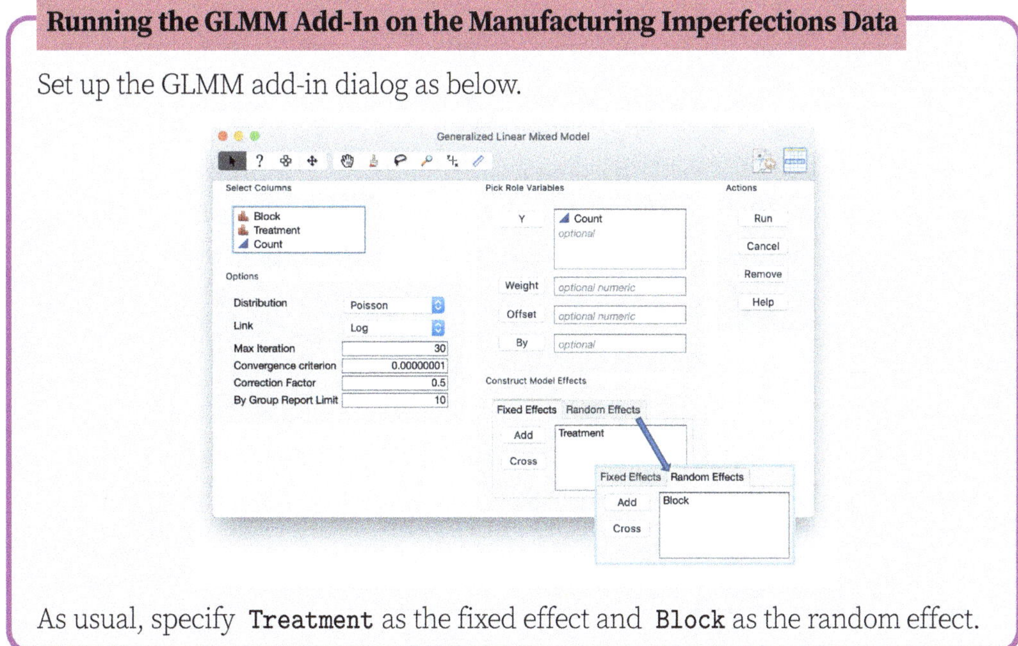

As usual, specify `Treatment` as the fixed effect and `Block` as the random effect.

Results are shown in Figure 9.7.

As shown the REML Variance Component Estimates table, Var Ratio column, block-to-block variance accounts for only around 7% of the total variance. The residual (within block) variability is much larger, and note that this is variability unexplained by blocks or treatments. It may represent inherent variability in the imperfection measurement process or perhaps additional covariates can be found to help explain it further.

The primary comparison of interest involves the treatment effects. The Least Squares Means plot reveals that Treatment 3 has substantially fewer imperfections on average than the others. It may represent a key breakthrough for improving product quality. Note interpretation here is on the log scale because that is the canonical link function for the Poisson distribution assumed in this GLMM.

Figure 9.7: GLMM Add-In Results for Manufacturing Imperfections Data

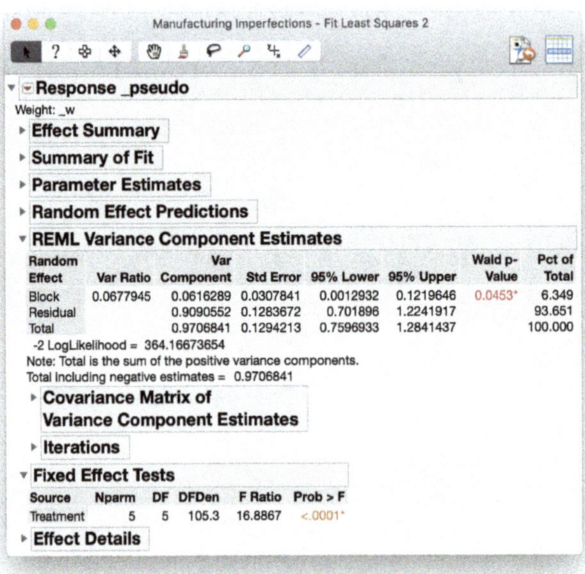

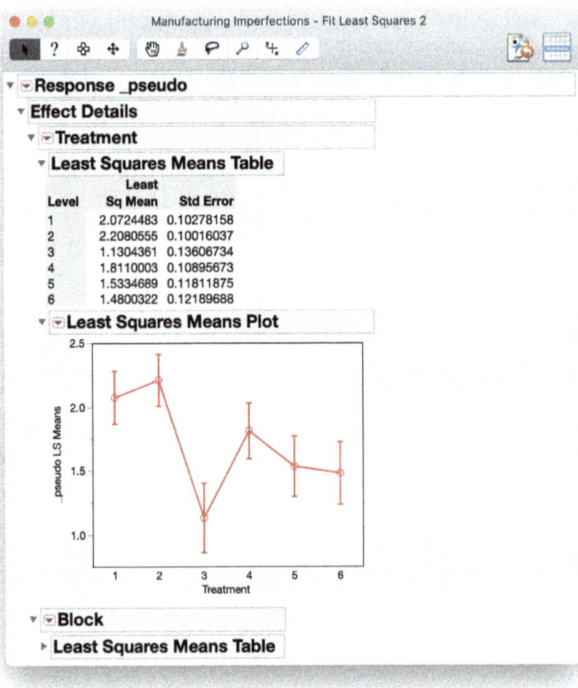

It may be more meaningful to transform results back to the original count scale. Care must be taken while doing this, as the exponential transformation is nonlinear. Certain statistics like standard errors are not directly transformable, and approximations like the delta method are often utilized in practice (Stroup et al., 2018). Confidence limits can be directly transformed.

Back-Transforming Confidence Limits

Suppose we want to compute confidence limits comparing Treatment 1 versus Treatment 3 on the original count scale. To do this, in the *Effect Details* section, click the red triangle next to *Treatment* > *LSMeans Contrast....* Click the + next to Treatment 1 and the - next to Treatment 3 to obtain a window as below. Click *Done*. The results of this comparison are for the difference between the two group means on the log scale.

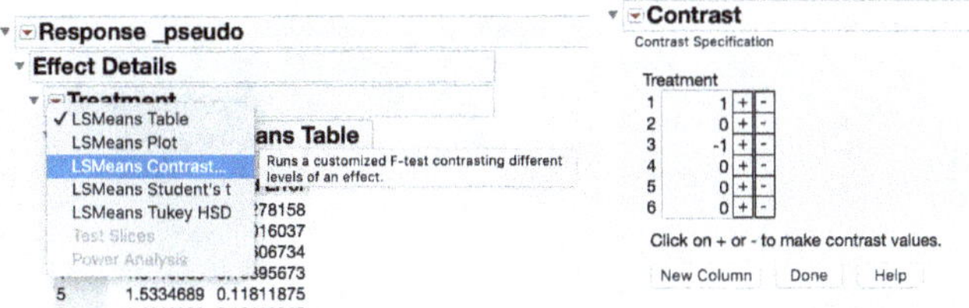

To transform this to a comparison on the original data scale, we need to back-transform the estimate and its confidence limits. To do this, expand the *Test Detail* box, right-click on the table, and choose *Make Into Data Table*.

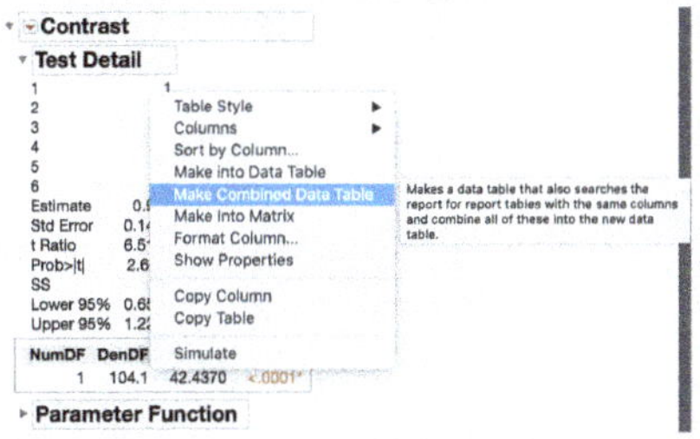

Back-Transforming Confidence Limits, Continued

In the resulting table, right-click on `Column 3` column > *New Formula Column* > *Transform* > *Exp*.

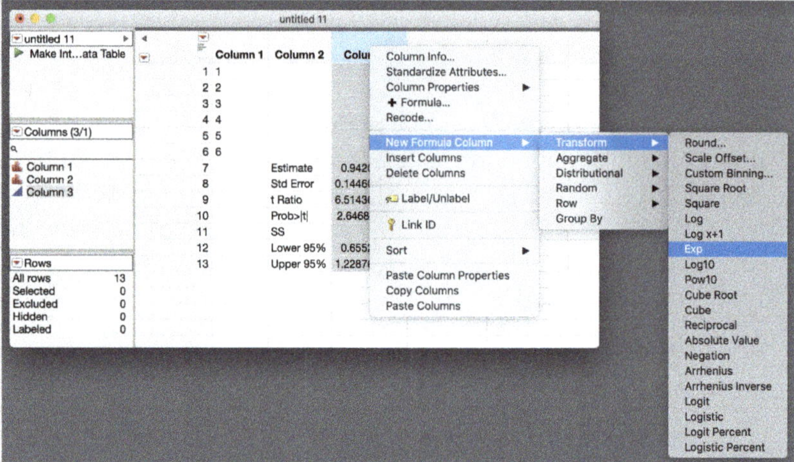

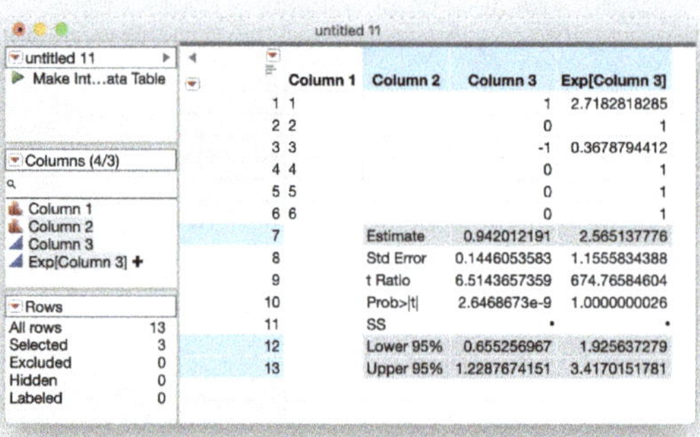

The column 3 Estimate (row 7) is the estimate difference between $\mu(\log(Count_{trt1})) - \mu(\log(Count_{trt3}))$. To return this to the original data scale, we exponentiate that difference, getting the value in column 4 (still row 7). However, note that this is no longer a difference of group means but instead is a ratio of the two groups, $\exp(A-B) = \exp(A)/\exp(B)$. In this example, we see that the mean of the logged values for Treatment 1 is 0.94 greater than the mean of the logged values of Treatment 3, and that the ratio of these values back on the original data scale is that Treatment 1 is 2.56 times Treatment 3.

The entries in the last two rows of the new column provide the approximate 95% confidence limits of (1.9,3.4). Treatment 3 is estimated to provide an improvement of 2-3 times fewer imperfections than Treatment 1.

9.6 Exercises

1. In the shrub coverage data, fit a standard normal linear mixed model using Weight_P as the response. How do results compare to those in the text with a binomial response? How do you explain similarities or differences?
2. In the shrub coverage data, convert the binomial response to a binary one by creating a stacked data table with 100 times more rows and a new response with values 0 or 1. Run the GLMM add-in on this stacked data. How do the results compare to those in the text? What are the reasons behind the similarities or differences?
3. In the salamander mating example, make a plot of the Least Squares Means of the fpop*mpop effect. Explain how the graph shows an interaction.
4. In the salamander mating example, stack the y_1, y_2, and y_3 responses into a single response y and create a new variable Trial with values 1,2, and 3. Run a mixed model on this stacked data and interpret the results. Should Trial be a fixed or random effect?
5. In the manufacturing imperfections data, run the Distribution script included with the JMP table–look for a green arrow near the upper left corner. Why is this considered a balanced incomplete block design? Click on one or more histogram bars in the output to see the distribution in the other factor to help verify your conclusion; create a screen shot as an example along with a written explanation.
6. In the manufacturing imperfections data, create a new response that is the log of count. One easy way is to select Count in the table, right-click > *New Formula Column* > *Transform* > *Log*. Run *Distribution* on both responses. How do they compare?
7. In the manufacturing imperfections data, run a standard linear mixed model using log(count) as the response. How do results compare to the GLMM analysis in the text?

Chapter 10

Mixed Models Amidst Modern Debates

In this chapter, we discuss mixed models in the context of several popular debates within the statistical and data science communities. It should provide you with some deeper insights regarding their relationship to various perspectives and how you can navigate and distinguish them.

10.1 Statistical Pragmatism

In this book we largely follow *statistical pragmatism* —the "big picture of statistical inference" approach advocated by Kass (2011). The following diagram (Figure 1 from that article) illustrates the main components and their interrelations.

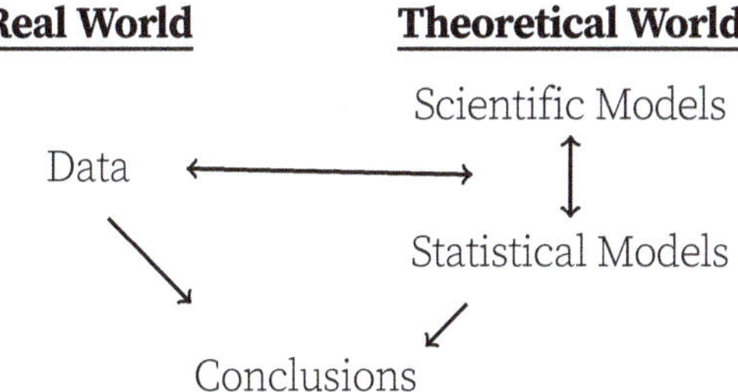

The key distinction is between the Real World, in which we observe data and conduct experiments, and the Theoretical World, in which we formulate Scientific and Statistical Models and their parameters. Specific mixed models are considered Statistical Models, and should (a) be based on a sound understanding of the accompanying scientific systems, and (b) reasonably represent and approximate the data-generating mechanisms being modeled, which in our case typically includes some form of covariance structure between observations. Mixed models can also be considered structural under appropriate causal assumptions; see Section 10.4. The real and theoretical worlds work in combination towards a general goal of reaching reliable conclusions using correlated data. Kass (2011) provides a more detailed breakdown in his Figure 4 (not shown), which includes Exploratory Data Analysis (EDA). JMP is an excellent tool for EDA, the primary techniques of which are readily and interactively available and complementary to mixed modeling.

The following sections adopt a pragmatic statistical viewpoint in discussing important major distinctions directly relevant to mixed modeling.

10.2 Fixed versus Random Effects

The distinction between fixed and random effects is powerful and nuanced and can be explained in several alternative ways. Since it is a central concept in this book, we provide some deeper discussion in this section.

If you've learned about analysis of variance (ANOVA) in an introductory or intermediate statistics course, you've probably (and perhaps unknowingly) been treating all of the categorical (nominal) predictor variables as fixed effects. There is an underlying assumption for fixed effects that the levels of that categorical variable are chosen purposefully. They are not just a random sample of the levels you could possibly have chosen. Instead, you care about those specific three schools, for example, or those eight varieties of wheat, or those two disease states (Yes and No, for example). In a fixed-effects ANOVA model, you use the traditional ANOVA table to fit the overall model, and then you follow this up with exploring the estimates or making other comparisons or contrasts between the mean response for the groups.

When you decide to treat a categorical predictor variable as a random effect, you are following the assumption that you did not choose the specific levels because you only care about those levels. Instead, you have a population of levels that are of interest to you. That is, you might only have the resources to test eight wheat varieties, but you would like to make inference to a full collection of wheat lines. You are assuming that, even if you did not actually take a truly random sample from the population of possible wheat varieties, your subset of eight varieties is reasonably similar to a truly random sample. This assumption that your categorical predictor variable contains only a random sample of the levels of interest changes the way this variable should be treated in the statistical model.

With random effects, instead of focusing on the differences in means between groups, we generally think about the variability they explain, and we often use that explained variance as a way to generalize the fixed effects means comparisons to extend across the entire population of possible values for the random effect. By explaining the variance that comes from the random effects, we can obtain confidence intervals on the estimates for the fixed effects that are true for any possible level of the random effect. In some cases, you may also have interest in the estimates for those "randomly" chosen groups of the random effect. These are possible to calculate, but estimation in this case can also be viewed as prediction of the realized values of the random effects. In such a case, we consider Best Linear Unbiased Predictions (BLUPs); see Chapter 5.

Alternative Definitions

Gelman (2005) and Gelman and Hill (2007) describe the following five alternative definitions of fixed and random effects:

1. Fixed effects are constant across individuals, and random effects vary. For example, in a growth study, a model with random intercepts a_i and fixed slope b corresponds to parallel lines for different individuals i, or the model $y_{it} = a_i + b_t$. Kreft and De Leeuw (1998) thus distinguish between fixed and random coefficients.
2. Effects are fixed if they are interesting in themselves or random if there is interest in the underlying population. Searle et al. (1992) explore this distinction in depth.
3. "When a sample exhausts the population, the corresponding variable is fixed; when the sample is a small (i.e., negligible) part of the population the corresponding variable is random." (Green and Tukey, 1960)
4. "If an effect is assumed to be a realized value of a random variable, it is called a random effect." (LaMotte, 1983)
5. Fixed effects are estimated using least squares (or, more generally, maximum likelihood) and random effects are estimated with shrinkage "linear unbiased prediction" in the terminology of Robinson (1991). This definition is standard in the multilevel modeling literature (see, for example, Snijders and Bosker (2012)) and in econometrics.

Gelman goes on to explain why he does not use the terms "fixed" and "random" effects due to differences in these definitions; he instead prefers "constant" and "varying."

Confusions, where they exist, appear to center around purely operational definitions (the literal statistical assumptions made in the model) versus more prescriptive definitions providing guidance for deciding how to apply these assumptions in practice. We do not see any serious conflicts here, rather different perspectives on the same issue. Our definitions of fixed and random effects from Chapter 1 are more operational in nature, and follow in the applied statistics tradition dating back to Eisenhart (1947), Henderson (1953), and Searle and Henderson (1961); see also Verbeke and Molenberghs (1997) and West et al. (2015). Google Scholar reveals that citations of articles and books on "mixed models" outnumber those from "multilevel models" by approximately ten times, and one would naturally expect some variations across this much larger literature. In addition, multilevel model approaches tend to begin with linear regression, whereas we feel a better starting point is blocking and ANOVA. Throughout this book, we provide examples of good ways to understand and apply fixed and random effects across a wide variety of experimental contexts.

Another way to understand fixed versus random effects is from a Bayesian perspective, in which all unknown parameters are assigned a prior probability distribution reflecting beliefs on how they relate to observed data. Under this paradigm, fixed effects in a linear model have a completely or effectively flat, uninformative prior. In contrast, random effects have an informative prior, in our case a normal distribution with zero mean and variance equal to a variance component. The random effect priors produce shrinkage as in alternative definition 5 above, since prior information is optimally combined with that in the data. Fixed effects can be viewed as random effects with infinite variance, and thus no shrinkage. A fully hierarchical Bayes approach goes further and assigns hyper-priors to the variance components, often using inverted gamma distributions as in Wolfinger and Kass (2000). The REML estimation approach that we use in this book can be considered an *empirical Bayes* approach, as the variance components are estimated directly from the data.

In general, your decision to declare effects fixed or random depends on the purpose of your analysis (Harville, 1992). While a typical purpose is tp assess differences between fixed effects and make inferences applicable to populations of random effects, you might instead wish to partition total variability into all known components. For this objective it can make sense to declare all effects to be random.

For the simple cell viability example from Chapter 1, experimental and inferential considerations point towards assigning Chemical as a fixed effect and Plate as a random effect. As design structure and data become more complex, the assignments can sometimes be difficult to make. The examples throughout this book and constructs like skeleton ANOVAs help to clarify many of the cases and were chosen to assist you in becoming proficient in making these decisions, as they are fundamental to becoming a skilled mixed modeler.

10.3 Frequentist versus Bayesian, p-values

A longstanding contention throughout the history of statistics and related fields concerns the meaning of probability when used in statistical models. *Frequentist* analyses typically view probability as *aleatoric*, describing distributions of real world events over a large number of hypothetical repetitions of an experiment, the prototypical example being the tossing of a coin. *Bayesian* analyses typically view probability as *epistemic*, a degree of belief and a formal mathematical way to describe uncertainty. The implications run deep and drive some strong differences in recommended ways to analyze experimental data, even in basic contexts like the examples that we consider in this book.

Both types of analyses have strengths and weaknesses. Bayesian methods often claim the theoretical high ground with a coherent methodology that optimally combines prior beliefs with data via Bayes Rule to produce posterior (updated) beliefs. An oftentimes practical obstacle is the necessity of specifying priors in terms of workable probability distributions, and justifying prior choices for any particular application and assessing sensitivity of results to them. Except for analytically conjugate cases, some form of Markov Chain Monte Carlo (MCMC) or related algorithm is typically required to simulate observations from the posterior distribution. By the way, JMP is a great tool for exploring results of MCMC posterior and related simulations; see Zink (2014).

Frequentist methods tend to be more objectively sharable and directly computable from data, and as a result have historically been the popular choice for statistical software packages (including JMP). The use of p-values in the context of frequentist Null Hypothesis Significance Testing has come under intense scrutiny recently; see Wasserstein et al. (2019) and all articles in that same issue of *The American Statistician*. Much of this targets abuses of the 0.05 rule of thumb like cherry-picking statistically significant results from a larger set, a practice known as *p-hacking*. The criticisms are part of a larger discussion in the scientific community to address problems of reproducibility. For sensible commentary and progress along these lines, with applications in the field of genomics, see research from the MAQC Society such as Shi et al. (2008), Scherer (2009), and SEQC/MAQC-III Consortium (2014).

Mixed models, as described herein, occupy somewhat of a middle ground between frequentism and Bayesianism. Output contains common frequentist statistics like p-values and confidence intervals but is based on Bayesian-style assumptions, especially the normal prior distributions on random effects. Results can usually be interpreted from a Bayesian perspective if desired. As mentioned in the previous section, mixed models can be properly viewed as an empirical Bayes methodology, see Efron (2010) and Carlin and Louis (2000).

Regarding p-values, it is important to clearly understand their definition: *The probability of observing a result as or more extreme than the current one, assuming that the null hypothesis is true.* Probability here is typically aleatoric. Note p-values depend on sample size and the experimental design that you use, and can be viewed as a mapping of signal-to-noise ratios (e.g. t-statistics) to a standardized scale based on tail areas of the assumed sampling distribution of the statistic under the null hypothesis. As such, and when interpreted properly, p-values are valuable statistics (Boos and Stefanski (2011)). Gibson (2020) provides a good general overview along with examples from pharmaceutical drug development and recommendations for moving forward.

Kass (2011) provides a general perspective and shows how statistical pragmatism moves beyond several key sticking points and differences between frequentist and Bayesian approaches. It is important to recognize that philosophical paradigms on how to best perform data analysis are not just limited to frequentism or Bayesianism. Others include eclecticism (Cox (1978), Wasserman (2004), Gelman and Hill (2007)), likelihood-based inference Royall (1997), Dempster-Shafer belief functions (Yager and Liu (2008)), fiducial inference (Hampel (2006), Hannig et al. (2016)), inferential modeling (Martin and Liu (2016), and causal modeling discussed in the next section. For example, Cisewski and

Hannig (2012) provide an interesting simulation-based approach for computing mixed model confidence intervals from a *fiducial* perspective, which can be viewed as a form of frequentist epistemic probability dating back to early writings of Fisher (Hampel (2006)). Another interesting approach is from Cheng et al. (2014), who derive plausibility intervals for heritabiltiy using an auxiliary variable derivation in an inferential modeling context, and Martin (2015) directly connects plausibility with p-values. Statistical pragmatism welcomes such approaches and is closely aligned with eclecticism, which by its nature attempts to amalgamate and reconcile best concepts and practices from differing philosophies.

Mayo (2018) provides a philosophically based review and journey through the major issues, and advocates for *severe testing*. The main principle here is to set up a series of difficult tests for your statistical models that will reveal flaws and weaknesses in their anticipated applications. Simulation is often a good tool for this; see Chapter 8. The blogs of Deborah Mayo, Andrew Gelman, and Larry Wasserman provide many informative and detailed exchanges, at times heated.

10.4 Causality versus Association

The well-worn maxim *Correlation is not Causation* provides a key warning accompanying most any type of data analysis. Given two variables X and Y, we say that X causes Y, with notation $X \rightarrow Y$, if manipulations of X produce changes in Y, all else held equal (*ceteris peribus*). In contrast, X and Y are *associated* if changes in one are observed to coincide with changes in the other. *Correlation* is the strength of linear relationship between two variables, although the term is often used more loosely and interchangeably with *association*. The distinction is subtle but very important. Causation is interventional and directional, whereas correlation is observational and symmetric. One famous case from medicine and epidemiology is described in (Correlation does not imply causation, 2019):

> For example, in a widely studied case, numerous epidemiological studies showed that women taking combined hormone replacement therapy (HRT) also had a lower-than-average incidence of coronary heart disease (CHD), leading doctors to propose that HRT was protective against CHD. But randomized controlled trials showed that HRT caused a small but statistically significant *increase* in risk of CHD. Re-analysis of the data from the epidemiological studies showed that women undertaking HRT were more likely to be from higher socio-economic groups (ABC1), with better-than-average diet and exercise regimens. The use of HRT and decreased incidence of coronary heart disease were coincident effects of a common cause (i.e. the benefits associated with a higher socioeconomic status), rather than a direct cause and effect, as had been supposed.

Here ABC1 was initially an *unmeasured confounder*, producing an association between

HRT and CHD. Measuring and properly adjusting for it is key in explaining why the relationship between HRT and CHD is associational and not causal. Either tracking down or properly accounting for unmeasured confounders is central to causal inference. The randomized controlled trials noted above accomplish this by recording and ideally balancing ABC1 and other potential confounders in their assignment of HRT. In general, careful laboratory protocols and expensive controlled experiments are a principled way to elucidate causal relationships in a system. Causal conclusions are usually preferred to associational ones, since the former typically provide deeper scientific insight than the latter and better aid in making scientific progress and wise policy decisions. In our context for this book, the risk is obtaining a strong association-based result from a mixed model analysis and slip unaware into dubious causal conclusions.

Reaching causal conclusions rides critically on initial assumptions. A relevant maxim from Cartwright (2007), and expounded upon by Pearl (2009), is *No Causes In, No Causes Out*. Another basic rule of thumb is if your experimental units exhibit a reasonable degree of consciousness — the prime example being human beings — then establishing causal conclusions tend to be much more difficult because such units can readily create unmeasured confounders. Hernán and Robins (2020) discuss the key role of *exchangability* of experimental units when making causal inferences, an assumption that we satisfy largely in this book by assuming independent and identically distributed (iid) random effects and errors along with inter-effect independence. Make sure these assumptions make sense in your mixed models before asserting causal conclusions, and be sure to include measures of uncertainty.

A historically rich, diverse, and growing set of methods exist for causal inference, especially in the econometrics, statistics, social science, and epidemiology literatures; refer to Hernán and Robins (2020), Pearl (2009), Heckman (2008), Imbens and Rubin (2015), White et al. (2009), Rosenbaum (2010), Morgan and Winship (2007), van der Laan and Rose (2011), Cartwright (2007), Sobel (1996), and Dawid (1979). These involve central concepts of confounding and counterfactuals and methods such as structural equation modeling, directed acyclic graphs (DAGs), potential outcomes, instrumental variables, g-estimation, matching, and propensity scoring. In this book, we do not cover these, but rather make the aforementioned basic exchangeability assumption. With appropriately controlled experimental designs, we can typically safely make this assumption (especially with sufficiently non-conscious experimental units), and thereby make causal statements from mixed models.

One justification of why we are typically in a good position to make causal statements in this book involves the linearity assumption of a normal mixed model. A fundamental precept of causality involves *counterfactual* (contrary to actual, hypothetical) questions such as *What would the response have been for a particular experimental unit had its treatment assignment been different?* For the cell viability example, this would be *What would have happened if we had swapped all of the chemical A and B assignments on half-plates?*

Assuming that a mixed model is a reasonably accurate approximation to the true data-generating mechanism, and that Chemical → Y and Plate → Y, we can use the fitted model to predict counterfactual outcomes by simply changing the treatment assignment as desired in the model prediction formula. We can also predict values for the actually observed responses in attempt to remove the effect of random residual error. We then estimate individual causal effects by taking the difference of these two predictions, a method known as g-estimation (Hernán and Robins, 2020). Because of linearity, everything cancels in the difference except the estimated treatment effect, which we have already estimated directly.

To illustrate with the cell viability data, fit a simple mixed model as in Section 1.2. Then from the red triangle next to *Response Y* in the output window, click > *Save Columns* > *Conditional Pred Formula*. This creates a new column in the JMP table containing a JMP Scripting Language (JSL) function of the fitted mixed model. Your table should appear as in Figure 10.1.

Figure 10.1: Cell Viability Data Table with Conditional Prediction Formula

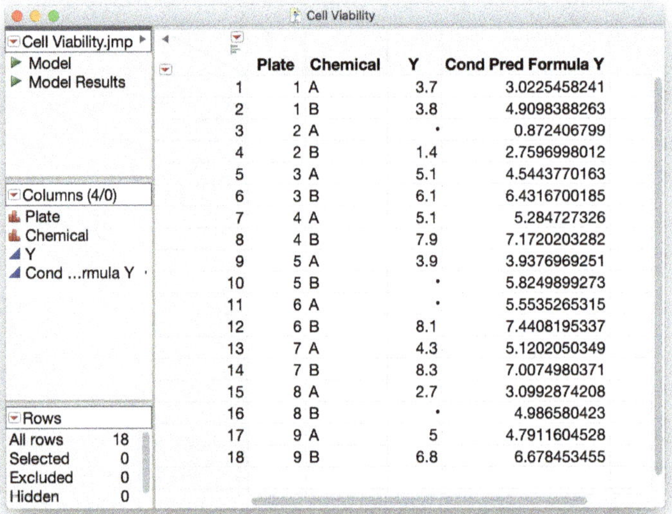

Cell Viability

Cell Viability.jmp
Model
Model Results

Columns (4/0)
Plate
Chemical
Y
Cond ...rmula Y

Rows
All rows 18
Selected 0
Excluded 0
Hidden 0

	Plate	Chemical	Y	Cond Pred Formula Y
1	1	A	3.7	3.0225458241
2	1	B	3.8	4.9098388263
3	2	A	•	0.872406799
4	2	B	1.4	2.7596998012
5	3	A	5.1	4.5443770163
6	3	B	6.1	6.4316700185
7	4	A	5.1	5.284727326
8	4	B	7.9	7.1720203282
9	5	A	3.9	3.9376969251
10	5	B	•	5.8249899273
11	6	A	•	5.5535265315
12	6	B	8.1	7.4408195337
13	7	A	4.3	5.1202050349
14	7	B	8.3	7.0074980371
15	8	A	2.7	3.0992874208
16	8	B	•	4.986580423
17	9	A	5	4.7911604528
18	9	B	6.8	6.678453455

Following the method above for estimating individual causal effects, we could make duplicates of all rows in the table and switch the Chemical value from A to B or B to A to create counterfactual predictions. Here this is unnecessary, as the values in the switched rows become identical to the already existing rows with opposite value of Chemical. Furthermore, differences between the B and A predictions within any one plate exactly equal $\hat{c}_2 - \hat{c}_1 = 2 * (0.94) = 1.88$. In this equation the notation $\hat{c}_1$ denotes the model-based estimate of c_1, and likewise for $\hat{c}_2$. Our mixed model assumes an identical causal effect of Chemical on Y for every plate, which is reasonable for this simple laboratory experiment.

10.5 Explanation versus Prediction

Shmueli (2010) provides a detailed dialectical comparison of two primary directions that you may take in your statistical inferences. The first is *to explain* what is going on in the data, moving towards a true understanding of the causal mechanisms in the scientific system under study. The second is *to predict* outcomes, an ability that can be very useful in current and future applications of your findings. Prediction is typically the main focus of machine learning and is also utilized in several causal inference techniques like the one at the end of the previous section and others like propensity scoring. Of course, if you truly understand a system, you can predict how it will behave, so in this sense at least the first direction has loftier ambitions than the second. This is also borne out in the desire to interpret black-box machine learning methods Lakkaraju et al. (2017).

Mixed models typically have straightforward linear components that lend themselves well to explanation, interpretation, and understanding. Most of the analyses in this book are of this nature, and we usually use familiar frequentist statistics such as t-ratios, p-values, confidence intervals, and estimates of variance components, combined with informative graphical displays.

You can also predict effectively with mixed models, using Best Linear Unbiased Predictions (BLUPs) of the random effects (see Chapter 5 and articles such as Robinson (1991)) along with estimates of the fixed effects parameters corresponding to a particular observation. As one common example, BLUPs are routinely used in plant and animal breeding for selecting the next round of crosses, a method dating back to Henderson (1953), one of the early pioneers of mixed model methodology. In JMP, such predictions are called *conditional*, as they condition on the predicted realized values of the random effects.

10.6 Randomized Experiments versus Observational Studies

When first encountering a new data set, an important question that you should immediately ask is: How were these data collected? A key distinction exists between data arising from a *designed experiment*, in which certain factors and conditions are explicitly controlled in attempt to eliminate or largely reduce the influence of any extraneous sources of variability, and *observational data*, which are collected from some naturally occurring "real world" system. Both types of data are immensely useful, but typically need to be handled differently, especially in terms of assumptions made for statistical modeling.

In this book, we focus primarily on data arising from designed experiments or data for which typical linear mixed model assumptions apply. This is not to say mixed models cannot be utilized with observational data (they indeed can), but designed experiments are typically a more natural match with mixed model techniques and concepts, and therefore a better milieu in which to learn them.

The distinction between experimental and observational data is not completely cut and dry. For example, you may be collecting data from a sample survey, which follows a planned survey design. Unlike a formally designed experiment, the subjects of a sample survey may not be randomly assigned to treatment groups according to a predetermined plan, but instead the survey design may address aspects of data collection such as how many subjects already possessing a certain trait (such as age group or disease status) should be collected. The key difference between a designed experiment and a sample survey is that in a designed experiment the treatment factors of interest are systematically assigned to the experimental units, whereas in a sample survey the treatment factors are not assigned to experimental units but instead are just observed. A key similarity for these two data sources is that there is some amount of predetermined design involved in the collection of the data. There is a rich history and wide variety of methods around survey sample data that we do not cover in this book; see the comprehensive collection of survey analysis procedures in SAS/STAT. Some direct connections and overlap with mixed model methods exist (e.g. sampling units, small area estimation, spatial smoothing).

A clinical study is a prevalent form of designed experiment in that the treatment is assigned to the subjects according to a specific protocol, usually involving inclusion/exclusion criteria, blocking, and randomization. A clinical study might also have some similarities to a sample survey in that some of the variables of interest cannot be assigned to the subjects but instead must simply be observed. In such cases, we must be careful with assumptions made, and these typically involve conditioning directly on the observed values of the covariates. A related phenomenon to guard against is *selection bias,* in which experimental circumstances make assignments of certain classes of individuals more likely and can create unwanted confounding. In general, a randomized clinical trial can be considered a designed experiment, and resulting data are typically quite amenable to mixed model analysis and causal inference.

A common example is a clinical trial that has two factors, Treatment (say involving an investigational therapeutic drug) and Center (a collection of hospitals at which screened patients are administered a level of Treatment). Treatment is naturally considered as a fixed effect and Center as a random effect, although several complexities can arise, including how to handle Treatment by Center *interactions* (the drug is performing differently at different hospitals).

Time series and spatial data are also very prevalent and involve factors that order the data in time and space, and hence cannot be randomized. These naturally impose a covariance structure that is a function of various difference or distance metrics. These types of data can often be well-handled by mixed models, especially when the relationships form small clusters.

10.7 Exercises

1. In the analysis at the end of Section 10.4, confirm that the estimated individual causal effect always equals 1.88 by taking the difference in conditional predictions within three different plates. Also, note the saved formula predicts values of Y that are missing in the original experiment. What assumptions make these predictions valid?
2. Research a topic from one of the sections of this chapter and write a brief report on your findings and how they relate to mixed modeling.

Appendix A

List of Examples Used in This Book

Chapter 1

Cell Viability — Imagine you are lab scientist studying the effect of two chemicals, A and B, on cell viability. You prepare nine plates of media with healthy cells growing on each, and then apply A and B to randomly assigned halves of each plate. After a suitable incubation period, you collect treated cells from the halves of each plate and perform an assay on each sample to compute a measurement Y of interest. Four of the samples are accidentally contaminated during processing and produce no assay results.

Chapter 2

Metal Bond Breaking — A fabrication company uses various metals (the treatment) as bonding agents to bond pieces of another composition material together. The company is interested in studying the pressure required to break the bonds. The experiment is conducted by taking ingots of the composition materials (seven in total) and subdividing them so that each treatment can be randomly assigned to a piece from each ingot.

Balanced Incomplete Blocks — A researcher has four blocks of material, and each block can be used for three runs. The researcher has four treatment levels to test. The treatments are randomized to the blocks so that each treatment shows up in three of the four blocks, subject to the constraint that each pair of treatments will be used in at least one block together.

In addition, we cover the following examples in the Exercises section at the end of this chapter.

Cell Viability — For the Cell Viability example from Chapter 1, we compare the effect of treating the Plates as fixed versus random.

Machine and Operator — The director of operations at a small manufacturing company suspects that the three machines in her machine shop are producing at different levels of quality. The company develops a continuous part quality rating scale, from 0 to 100, and she wants to rate 18 randomly selected parts produced by each machine (for a total of 54 parts). Six machine operators will assess the ratings of the parts. Each operator will rate three parts from each of the three machines. The order in which the parts are chosen and assigned to the operators for rating is random.

Cotton — An experimenter is comparing the yield, in pounds of seed cotton per plot, of fifteen field treatments. The experimenter only has fifteen strips of field to use for the experiment, and each strip can be divided into four plots on which to apply different treatments.

Chapter 3

Tensile Strength — A fabric manufacturer wants to test the tensile strength of a fabric after washing under various conditions. She has three machines available each with a setting for temperature, hot and cold, and for water level, high and low. It is possible that both temperature and water level affect the strength of the fabric.

Greenhouse — A plant researcher has two plant varieties and a pesticide meant to protect the plants against disease. In the greenhouse, the pesticide can only be applied to large sections of the benches. The bench sections can hold multiple plants. The researcher wants to identify the best plant and pesticide combination for disease resistance.

Semiconductor — A semiconductor engineer has several process conditions that affect resistance on the wafers produced. The position of the chips on the wafer can also affect resistance. The engineer wants to minimize resistance and understand the effects of process condition and position.

Chapter 4

Fabric Shrinkage — A fabric manufacturer needs to test four new materials to be used in permanent press garments. The heat chamber used to test fabrics has four positions. Each fabric should be tested under each position, and due to time constraints, the manufacturer is limited to four runs. Fabric shrinkage is the response of interest.

Mouse Condition — A trial involving laboratory mice investigates the effects of four housing conditions (CONDITION 1, 2, 3, and 4) and three feeding regimens (DIET "restricted," "normal," and "supplemental") on weight gain in mice. As described in Example 5.7 in Stroup et al. (2018), two housing conditions can be accommodated in one CAGE unit, and within each housing-condition-in-a-caging-unit, mice can be separated into the three distinct diet groups.

In the Exercises section at the end of this chapter, we also cover the following example.

Semiconductor — In an extension of the semiconductor experiment introduced in Chapter 3, the engineer again has several process conditions that could affect resistance on the wafers produced. The position of the chips on the wafer can also affect resistance. There is also potential variability in the process due to the position (another blocking factor). The engineer wants to minimize resistance and understand whether position of the chip on the wafer affects resistance.

Chapter 5

Stability Trial — Drug developers perform a stability trial to determine optimal shelf life of a pharmaceutical. The batches used to measure the strength over time are randomly selected from possible batches created.

Student Achievement — Eighth grade students take mathematics achievement tests. Researchers are interested in not only any overall trend among all students but also trends within individual classrooms.

And, in the Exercises section:

Winter Wheat — The data table `winterwheat.jmp` contains data from ten varieties of wheat that were randomly selected from the population of varieties of hard red winter wheat. The varieties were randomized to sixty, 1-acre plots in a field for six replicates for each variety. It was thought that the preplanting moisture of the soil could influence the germination rate and hence the eventual yield. So, the amount of moisture in the top 36 inches of soil was determined from a core sample at the middle of each plot. The response is bushels per acre, `yield`, and the covariate is the amount of preplanting moisture, `moist`.

Chapter 6

Respiratory Ability — A pharmaceutical company examines effects of three drugs on respiratory ability of asthma patients. Treatments include a standard drug (A), a test drug (C), and a placebo (P). The drugs are randomly assigned to 24 patients each. The assigned treatment is administered to each patient, and a standard measure of respiratory ability called FEV1 is measured hourly for 8 hours following treatment. FEV1 is also measured immediately prior to administration of the drugs.

In addition, we cover the following examples in the Exercises section at the end of this chapter.

Cholesterol Study — In the `Cholesterol Stacked.jmp` data table, there are five subjects in four treatment groups, with measurements taken in the morning and afternoon, once a month, for three months. The goal is to fit a model for the response, Y, based on the Treatment, Month, and AM/PM as full factorial effects.

Tiretread — Four characteristics (Abrasion, Modulus, Elongation, and Hardness) were measured on each of twenty tires. The tires consist of various amounts of Silica, Silane, and Sulfur. The experiment was arranged as a Response Surface design in order to explore the main treatment effects, the two-way treatment interactions, and the treatment quadratic effects.

Chapter 7

Hazardous Waste — The investigator for an environmental study wants to evaluate water drainage at a potential hazardous waste disposal site. It is believed that water movement is affected by the thickness of a layer of salt, a geological feature of the area. Samples taken closer together are more alike than those farther apart.

Wheat Trial — A researcher wants to compare fifty-six varieties of wheat. A blocked design is frequently used for this type of experiment, so four complete blocks were planted in a field. Initial statistical results from the RCB trial do not "make sense" to the subject-matter expert. A different model may produce sensible conclusions.

The Exercises section at the end of this chapter includes the following example.

Seed Trial — The experiment was similar to the Alliance wheat trial of the second example. The variable `Rep` is the complete block with 3 complete blocks of the 48 treatments, `Trt`. The variable `Block` divides the complete blocks into incomplete blocks for a BIB design. The location of each experimental unit in the 12×12 grid was also recorded as `Lat` and `Lng`. The response was yield and recorded as `Y`. Higher yield is desirable.

Chapter 8

Semiconductor Experiment — Semiconductor researchers have process conditions and finishing treatments that can interact in affecting performance of the chip when complete. They have a new design that they need to investigate how it will perform given what they already know about their chip manufacturing process.

Wheat Trial — The wheat farmers need to know whether the statistical model that they are using to analyze their data produces confidence intervals that are actually 95% intervals as they state. They have heard that sometimes a procedure only has coverage of 90% or less despite stating 95%. They want to be confident about their confidence.

All the Statistics — A researcher wants to investigate the properties of several different statistics produced in analyzing a randomized complete block design. They would like a compact way to study them all without having to run separate studies for each individual statistic.

Chapter 9

Binomial — The **Shrub Coverage** example from Parks Canada measures the number of sampling points out of 100 that contain species from a plant functional group during a given year. A "success" in this case is the presence of the functional group, and a "trial" is the examination of a plot for that type of plant.

Binary — (also known as *Bernouilli*). The **Salamander Mating** example is of this type. For these data, the response is an indicator of whether mating occurred in that experimental group. Related experimental features are the day on which the experiment occurred, the population group from which the females were taken (of two groups), the population group from which the males were taken (of the same two groups), and the numbers of females and males present in the experimental group.

Count — The **Manufacturing Imperfections** example represents a common type of data from industrial quality control. Here the counts are the number of defects in a manufactured part, and treatments represent different modes of production.

Chapter 10

Cell Viability — We revisit the Cell Viability example from Chapter 1 a final time to investigate possible causal claims between the treatments and response.

Bibliography

Bolker, Benjamin M., Mollie E. Brooks, Connie J. Clark, Shane W. Geange, John R. Poulsen, M. Henry H. Stevens, and Jada-Simone S. White. 2009. *Generalized linear mixed models: a practical guide for ecology and evolution*, Trends in Ecology & Evolution **24**, no. 3, 127 -135.

Boos, Dennis D. and Leonard A. Stefanski. 2011. *P-value precision and reproducibility*, The American Statistician **65**, no. 4, 213–221, available at https://doi.org/10.1198/tas.2011.10129.

Breslow, N.E. and D.G. Clayton. 1993. *Approximate inference in generalized linear mixed models*, Journal of the American Statistical Association **88**, no. 421, 9–25.

Brown, H and R Prescott. 2015. *Applied mixed models in medicine*, 3rd ed., Statistics in Practice, Wiley, Hoboken, NJ.

Carlin, Bradley P. and Thomas A. Louis. 2000. *Empirical bayes: Past, present and future*, Journal of the American Statistical Association **95**, no. 452, 1286–1289.

Cartwright, Nancy. 2007. *Hunting causes and using them: Approaches in philosophy and economics*, Cambridge University Press, New York.

Cheng, Qianshun, Xu Gao, and Ryan Martin. 2014. *Exact prior-free probabilistic inference on the heritability coefficient in a linear mixed model*, Electronic Journal of Statistics **8**, no. 2, 3062–3076.

Cisewski, Jessi and Jan Hannig. 2012. *Generalized fiducial inference for normal linear mixed models*, The Annals of Statistics **40**, no. 4, 2102–2127.

Cochran, WG and GM Cox. 1957. *Experimental designs*, Wiley, New York.

______. 1992. *Experimental designs, 2nd ed.*, Wiley, New York.

Correlation does not imply causation. 2019. *Correlation does not imply causation — Wikipedia, the free encyclopedia*. [Online; accessed 18-May-2019].

Cox, D. R. 1978. *Foundations of statistical inference: The case for eclecticism*, Australian Journal of Statistics **20**, no. 1, 43–59, available at https://onlinelibrary.wiley.com/doi/pdf/10.1111/j.1467-842X.1978.tb01094.x.

Cressie, NAC. 1991. *Statistics for spatial data*, Wiley, New York.

Crowder, M.J. and D.J. Hand. 1990. *Analysis of repeated measures*, Chapman & Hall/CRC Monographs on Statistics & Applied Probability, Chapman & Hall, New York.

Dawid, AP. 1979. *Conditional independence in statistical theory*, Journal of the Royal Statistical Society. Series B (Methodological) **41**, no. 1, 1–31.

Derringer, George and Ronald Suich. 1980. *Simultaneous optimization of several response variables*, Journal of Quality Technology **12**, no. 4, 214–219.

Dong, M. 2020. *Generalized linear mixed model add-in*, available at https://community.jmp.com/t5/JMP-Add-Ins/Generalized-Linear-Mixed-Model-Add-in/ta-p/284627.

Dua, Dheeru and Casey Graff. 2017. *UCI machine learning repository*, University of California, Irvine, School of Information and Computer Science.

Efron, Bradley. 2010. *Large-scale inference: Empirical bayes methods for estimation, testing, and prediction*, Institute of Mathematical Statistics Monographs, Cambridge University Press, New York.

Eisenhart, C. 1947. *The assumptions underlying the analysis of variance*, Biometrics **3**, no. 1, 1–21.

El-Saba, Muhammad H, 2015, *Measurement of the semiconductor parameters*, In Electronic engineering materials & nanotechnology, pp. 931–1050.

Federer, WT. 1955. *Experimental design: Theory and application*, Macmillan, New York and London.

Fisher, RA. 1925. *Statistical methods for research workers*, Oliver and Boyd, Edinburgh.

Gelman, A. 2005. *Why i don't use the term "fixed and random effects"*, available at https://statmodeling.stat.columbia.edu/2005/01/25/why_i_dont_use/.

Gelman, A and J Hill. 2007. *Data analysis using regression and multilevel/hierarchical models*, Cambridge University Press, New York.

Gibbs, P and K Kiernan, 2020, *Simulating data for complex linear models*, In Proceedings of the SAS global forum 2020 conference, SAS Institute, Inc., Cary, NC.

Gibson, Eric W. 2020. *The role of p-values in judging the strength of evidence and realistic replication expectations*, Statistics in Biopharmaceutical Research **13**, no. 1, 6–18, available at https://doi.org/10.1080/19466315.2020.1724560.

Goldstein, H. 1987. *Multilevel models in educational and social research*, Oxford University Press, New York.

Goos, P and B Jones. 2011. *Optimal design of experiments: A case study approach*, Wiley, Hoboken, NJ.

Green, BF and JW Tukey. 1960. *Complex analyses of variance: General problems*, Psychometrika **25**, no. 2, 127–152.

Hampel, F., 2006, *The proper fiducial argument*, In General theory of information transfer and combinatorics (Rudolf Ahlswede, Lars Bäumer, Ning Cai, Harout Aydinian, Vladimir Blinovsky, Christian Deppe, and Haik Mashurian, eds.), pp. 512–526, Springer, Berlin.

Hannig, Jan, Hari Iyer, Randy C. S. Lai, and Thomas C. M. Lee. 2016. *Generalized fiducial inference: A review and new results*, Journal of the American Statistical Association **111**, no. 515, 1346–1361, available at https://doi.org/10.1080/01621459.2016.1165102.

Harville, DA. 1976. *Extension of the Gauss-Markov theorem to include the estimation of random effects*, The Annals of Statistics **4**, no. 2, 384–395.

______. 1992. *personal communication.*

Heckman, JJ. 2008. *Econometric causality*, International Statistical Review **76**, no. 1, 1–27, available at https://onlinelibrary.wiley.com/doi/pdf/10.1111/j.1751-5823.2007.00024.x.

Henderson, CR. 1953. *Estimation of variance and covariance components*, Biometrics **9**, no. 2, 226–252.

______, 1963, *Selection index and expected genetic advance*, In Statistical genetics and plant breeding (W.D. Hanson and H.F. Robinson, eds.), vol. Publication No. 982, pp. 141–163, The National Academies Press, Washington, DC.

Hernán, MA and JM Robins. 2020. *Causal inference: What if*, Chapman & Hall / CRC, Boca Raton, FL.

Imbens, GW and DB Rubin. 2015. *Causal inference for statistics, social, and biomedical sciences: An introduction*, Cambridge University Press, New York.

International Conference on Harmonisation of Technical Requirements for Registration of Pharmaceuticals for Human Use. 2003. *ICH harmonised tripartite guideline: Evaluation for stability data Q1E.*

Isaaks, EH and RM Srivastava. 1989. *An introduction to applied geostatistics,* Oxford University Press, New York.

Journel, AG and CJ Huijbregts. 1978. *Mining geostatistics,* Academic Press, New York.

Kass, RE. 2011. *Statistical inference: The big picture (with rejoinder),* Statistical Science **26**, no. 1, 1–9.

Kenward, MG and JH Roger. 1997. *Small sample inference for fixed effects from restricted maximum likelihood,* Biometrics **53**, no. 3, 983–997.

Kreft, I and J De Leeuw. 1998. *Introducing multilevel modeling,* Sage, Thousand Oaks, CA.

Kreft, IGG, J De Leeuw, and R Van Der Leeden. 1994. *Review of five multilevel analysis programs: BMDP-5V, GENMOD, HLM, ML3, VARCL,* The American Statistician **48**, no. 4, 324–335.

Laird, NM and JH Ware. 1982. *Random-effects models for longitudinal data,* Biometrics **38**, no. 4, 963–974.

Lakkaraju, Himabindu, Ece Kamar, Rich Caruana, and Jure Leskovec. 2017. *Interpretable and explorable approximations of black box models.* Paper presented at the 4th Workshop on Fairness, Accountability, and Transparency in Machine Learning, Special Interest Group on Knowledge Discovery and Data Mining (SIGKDD).

LaMotte, LR, 1983, *Fixed-, random-, and mixed-effects models,* In Encyclopedia of statistical sciences, vol. 3, pp. 137–141, Wiley, New York.

Leung, Kwok. 2011. *Presenting post hoc hypotheses as a priori: Ethical and theoretical issues,* Management and Organization Review **7**, no. 3, 471–479.

Littell, RC, GA Milliken, WW Stroup, and RD Wolfinger. 1996. *SAS system for mixed models,* 1st ed., SAS Institute, Inc., Cary, NC.

Littell, RC, GA Milliken, WW Stroup, RD Wolfinger, and O Schabenberger. 2006. *SAS for mixed models,* 2nd ed., SAS Institute, Inc., Cary, NC.

Littell, RC, J Pendergast, and R Natarajan. 2000. *Tutorial in biostatistics: Modelling covariance structure in the analysis of repeated measures data,* Statistics in Medicine **19**, no. 13, 1793–1819.

Littell, RC, WW Stroup, and RJ Freund. 2002. *SAS for linear models,* 4th ed., SAS Institute Inc., Cary, NC.

Martin, R. and C. Liu. 2016. *Inferential models: Reasoning with uncertainty,* Chapman & Hall/CRC Monographs on Statistics and Applied Probability, CRC Press, Boca Raton, FL.

Martin, Ryan. 2015. *Plausibility functions and exact frequentist inference,* Journal of the American Statistical Association **110**, no. 512, 1552–1561, available at `https://doi.org/10.1080/01621459.2014.983232`.

Mayo, D.G. 2018. *Statistical inference as severe testing: How to get beyond the statistics wars,* Cambridge University Press, New York.

McCullagh, P and J.A. Nelder. 1989. *Generalized linear models,* 2nd ed., Chapman & Hall, New York.

McLean, Robert A, William L Sanders, and Walter W Stroup. 1991. *A unified approach to mixed linear models,* The American Statistician **45**, no. 1, 54–64.

Milliken, George A and Dallas E Johnson. 2009. *Analysis of messy data, volume 1: designed experiments,* Chapman & Hall/CRC, Boca Raton, FL.

Montgomery, Douglas C. 2005. *Design and analysis of experiments, 6th ed.,* Wiley, Hoboken, NJ.

Morgan, SL and C Winship. 2007. *Counterfactuals and causal inference: Methods and principles for social research,* Analytical Methods for Social Research, Cambridge University Press, New York.

Parris, J. and R. Hummel. 2021a. *FDR multiple reports add-in,* available at `https://community.jmp.com/t5/JMP-Add-Ins/`.

______. 2021b. *Repeated measures cov/corr diagnostics add-in,* available at `https://community.jmp.com/t5/JMP-Add-Ins/`.

Pearl, J. 2009. *Causality: Models, reasoning, and inference*, 2nd ed., Cambridge University Press, New York.

Peters, J., D. Janzing, and B. Schölkopf. 2017. *Elements of causal inference: Foundations and learning algorithms*, Adaptive Computation and Machine Learning series, MIT Press, Cambridge, MA.

Raudenbush, SW and AS Bryk. 2002. *Hierarchical linear models: Applications and data analysis methods*, Sage Publications, Thousand Oaks, CA.

Robinson, GK. 1991. *That BLUP is a good thing: The estimation of random effects (with rejoinder)*, Statistical Science **6**, no. 1, 15–32.

Rosenbaum, PR. 2010. *Design of observational studies*, Springer Series in Statistics, Springer, New York.

Royall, RM. 1997. *Statistical evidence: A likelihood paradigm*, Chapman & Hall/CRC, Boca Raton, FL.

Rutter, CM and RM Elashoff. 1994. *Analysis of longitudinal data: Random coefficient regression modelling*, Statistics in Medicine **13**, no. 12, 1211–1231.

Sall, J. 2014. *Nonlinear degradation mixed model*, available at https://community.jmp.com/t5/JMP-Scripts/Nonlinear-Degradation-Mixed-Model/ta-p/21367.

Samson, Claude. 2021. *Proportion of canopy layer occupied by shrubs, lichens and graminoids - data extracted from the auyuittuq national park's plant community monitoring data 2010-2019*, Parks Canada, Ecological Monitoring Division.

SAS Institute Inc. 2019a. *JMP documentation library*, available at https://jmp.com/help/.

______. 2019b. *Using SAS from JMP*. Last modified: 2019-11. Accessed: 2020-08-10.

Schabenberger, O and CA Gotway. 2005. *Statistical methods for spatial data analysis*, Chapman & Hall/CRC, Boca Raton, FL.

Schall, R. 1991. *Estimation in generalized linear models with random effects*, Biometrika **78(4)**, 719–727.

Scherer, A. 2009. *Batch effects and noise in microarray experiments: Sources and solutions*, Wiley Series in Probability and Statistics, John Wiley & Sons, Chichester, UK.

Searle, SR, G Casella, and CE McCulloch. 1992. *Variance components*, Wiley, New York.

Searle, SR and CR Henderson. 1961. *Computing procedures for estimating components of variance in the two-way classification, mixed model*, Biometrics **17**, no. 4, 607–616.

SEQC/MAQC-III Consortium, The. 2014. *A comprehensive assessment of rna-seq accuracy, reproducibility and information content by the sequencing quality control consortium*, Nature Biotechnology **32(9)**, 903–914, available at https://www.ncbi.nlm.nih.gov/pmc/articles/PMC4321899/.

Shi, L, WD Jones, and RV Jensen et al. 2008. *The balance of reproducibility, sensitivity, and specificity of lists of differentially expressed genes in microarray studies*, BMC Bioinformatics **9**, available at https://bmcbioinformatics.biomedcentral.com/articles/10.1186/1471-2105-9-S9-S10.

Shmueli, G. 2010. *To explain or to predict?*, Statistical Science **25**, no. 3, 289–310.

Singer, JD. 1998. *Using SAS PROC MIXED to fit multilevel models, hierarchical models, and individual growth models*, Journal of Educational and Behavioral Statistics **23**, no. 4, 323–355.

Snijders, TAB and RJ Bosker. 2012. *Multilevel analysis: An introduction to basic and advanced multilevel modeling*, 2nd ed., Sage, Los Angeles.

Sobel, ME. 1996. *An introduction to causal inference*, Sociological Methods & Research **24**, no. 3, 353–379, available at https://doi.org/10.1177/0049124196024003004.

Stroup, Walt and Elizabeth Claassen. 2020. *Pseudo-likelihood or quadrature? what we thought we knew, what we think we know, and what we are still trying to figure out*, Journal of Agricultural, Biological and Environmental Statistics **25**, no. 4, 639–656.

Stroup, Walter W. 2012. *Generalized linear mixed models: modern concepts, methods and applications*, CRC Press, Taylor & Francis Group, Boca Raton, FL.

Stroup, WW. 2002. *Power analysis based on spatial effects mixed models: A tool for comparing design and analysis strategies in the presence of spatial variability*, Journal of Agricultural, Biological, and Environmental Statistics **7**, no. 4, 491–511.

Stroup, WW, PS Baenziger, and DK Mulitze. 1994. *Removing spatial variation from wheat yield trials: a comparison of methods*, Crop Science **34**, no. 1, 62–66.

Stroup, WW, GA Milliken, EA Claassen, and RD Wolfinger. 2018. *SAS for mixed models: Introduction and basic applications*, SAS Institute Inc., Cary, NC.

Stroup, WW and M Quinlan, 2016, *Statistical considerations for stability and the estimation of shelf life*, In Nonclinical statistics for pharmaceutical and biotechnology industries (Lanzu Zhang, ed.), pp. 575–604, Springer International Publishing, Cham, Switzerland.

van der Laan, M.J. and S. Rose. 2011. *Targeted learning: Causal inference for observational and experimental data*, Springer Series in Statistics, Springer, New York.

Verbeke, Geert and Geert Molenberghs. 1997. *Linear mixed models in practice: A SAS-oriented approach*, Springer, New York.

Wackerly, DD, W Mendenhall III, and RL Scheaffer. 1996. *Mathematical statistics with applications*, 5th ed., Duxbury Press, Belmont, CA.

Wasserman, L. 2004. *All of statistics: A concise course in statistical inference*, Springer Texts in Statistics, Springer, New York.

Wasserstein, Ronald L., Allen L. Schirm, and Nicole A. Lazar. 2019. *Moving to a world beyond p < 0.05*, The American Statistician **73**, no. supl, 1–19.

West, B.T., K.B. Welch, and A.T. Galecki. 2015. *Linear mixed models: A practical guide using statistical software, second edition*, CRC Press, Taylor & Francis Group, Boca Raton, FL.

Westfall, PH and AL Arias. 2020. *Understanding regression analysis: A conditional distribution approach*, CRC Press, Boca Raton, FL.

White, Halbert, Karim Chalak, and Xun Lu, 2009, *Linking Granger causality and the Pearl causal model with settable systems*, In NIPSMINI '09: Proceedings of the 12th international conference on neural information processing systems (NIPS) mini-symposium on causality in time series, p. 1–29.

Wicklin, R. 2013. *Simulating data with SAS®*, SAS Institute, Inc., Cary, NC.

Wolfinger, RD. 1996. *Heterogeneous variance covariance structures for repeated measures*, Journal of Agricultural, Biological, and Environmental Statistics **1(2)**, 205–230.

Wolfinger, RD and M O'Connell. 1993. *Generalized linear mixed models: A pseudo-likelihood approach*, Journal of Statistical Computation and Simulation **48**, no. 3–4, 233–243.

Wolfinger, Russell D. and Robert E. Kass. 2000. *Nonconjugate bayesian analysis of variance component models*, Biometrics **56**, no. 3, 768–774, available at https://onlinelibrary.wiley.com/doi/pdf/10.1111/j.0006-341X.2000.00768.x.

Yager, Ronald R. and Liping Liu. 2008. *Classic works of the Dempster-Shafer theory of belief functions*, 1st ed., Springer, New York.

Yates, Frank. 1935. *Complex experiments*, Supplement to the Journal of the Royal Statistical Society **2**, no. 2, 181–223.

Zeger, SL, KY Liang, and Alberg PS. 1988. *Models for longitudinal data: A generalized estimating equation approach*, Biometrics **44**, no. 4, 1049–1060.

Zink, R. 2014. *MCMC diagnostics add-in*, available at https://community.jmp.com/t5/JMP-Add-Ins/MCMC-Diagnostics-Add-In/ta-p/21500.

www.ingramcontent.com/pod-product-compliance
Lightning Source LLC
LaVergne TN
LVHW080114070326
832902LV00015B/2584
* 9 7 8 1 9 5 1 6 8 4 0 2 0 *